Modelling and Simulations for Tourism and Hospitality

TOURISM ESSENTIALS

Series Editors: **Chris Cooper** *(Leeds Beckett University, UK)*, **C. Michael Hall** *(University of Canterbury, New Zealand)* and **Dallen J. Timothy** *(Arizona State University, USA)*

Tourism Essentials is a dynamic new book series of short accessible volumes focusing on a specific area of tourism studies. It aims to present cutting-edge research on significant and emerging topics in tourism, providing a concise overview of the field as well as examining the key issues and future research possibilities. This series aims to create a new generation of tourism authors by encouraging young researchers as well as more established academics. The books will provide insight into the latest perspectives in tourism studies and will be an essential resource for postgraduate students and researchers.

All books in this series are externally peer-reviewed.

Full details of all the books in this series and of all our other publications can be found on http://www.channelviewpublications.com, or by writing to Channel View Publications, St Nicholas House, 31-34 High Street, Bristol BS1 2AW, UK.

TOURISM ESSENTIALS: 6

Modelling and Simulations for Tourism and Hospitality

An Introduction

Jacopo A. Baggio and Rodolfo Baggio

CHANNEL VIEW PUBLICATIONS
Bristol • Blue Ridge Summit

To a woman (mother and wife) who inspired our lives and will not read this book.

DOI https://doi.org/10.21832/BAGGIO7420
Library of Congress Cataloging in Publication Data
A catalog record for this book is available from the Library of Congress.

Library of Congress Control Number: 2019035822

British Library Cataloguing in Publication Data
A catalogue entry for this book is available from the British Library.

ISBN-13: 978-1-84541-742-0 (hbk)
ISBN-13: 978-1-84541-741-3 (pbk)

Channel View Publications
UK: St Nicholas House, 31-34 High Street, Bristol BS1 2AW, UK.
USA: NBN, Blue Ridge Summit, PA, USA.

Website: www.channelviewpublications.com
Twitter: Channel_View
Facebook: https://www.facebook.com/channelviewpublications
Blog: www.channelviewpublications.wordpress.com

The policy of Multilingual Matters/Channel View Publications is to use papers that are natural, renewable and recyclable products, made from wood grown in sustainable forests. In the manufacturing process of our books, and to further support our policy, preference is given to printers that have FSC and PEFC Chain of Custody certification. The FSC and/or PEFC logos will appear on those books where full certification has been granted to the printer concerned.

Typeset by Deanta Global Publishing Services, Chennai, India.

Contents

Figures and Tables

Figures

Tables

Introduction

Essentially, all models are wrong, but some are useful. This famous aphorism, attributed to the renowned statistician George Box, seems to be a standard beginning for any book on modelling and simulation. Despite its resemblance to a joke, the aphorism contains an essence of truth: that this is more an art than a science.

Even though a vast array of scholars and practitioners of all disciplines have produced and implemented a wide number of methods and tools, the assembly and operation of a set of techniques that can supply meaningful answers to a question is a task that requires not only good technical competences but, even more importantly, good experience and a familiarity with many, often not completely clear, concepts.

Moreover, given the contemporary landscape, it would be quite uncommon for one individual to have all the skills and the practice required. Modelling is thus a multidisciplinary endeavour, in which diverse know-hows need to be composed in a smooth and harmonious way.

The tourism domain has progressed considerably in its ability to understand the phenomenon and the components that make this one of the most interesting and fascinating areas. However, the tourism researcher's or analyst's toolbox has seen little improvement. Most of the works published use rather traditional approaches and methods and concentrate more on a wide number of detailed investigations, often losing sight of the larger issues.

With the recognition, nowadays well grounded, of the complex nature of the phenomenon, of the systems involved and of their relationships, internal or external, there is a need, well expressed on several occasions, to proceed towards different perspectives. Today, this is also imposed by the incredible advances in information and communication technologies that have made available a wealth of means, algorithms and techniques that can be usefully employed for the purpose.

With this work, we try to raise the awareness of tourism and hospitality researchers by providing an essential introduction to the most popular methods useful for modelling and simulating systems and phenomena of interest for those who deal with the intricate and complex world of tourism.

We are well aware that we only scratch the surface of the domain, and we do not pretend to provide a fully fledged manual, as this would be an almost impossible endeavour. Some of the techniques described here require a highly complicated and sophisticated technical background and the interested reader can refer to specific textbooks on such techniques. Here, we have chosen to focus on the main concepts and avoid, as much as possible, the technicalities and descriptions of the nuts and bolts of the methods introduced.

We think that a compact and documented work providing an overview of techniques and methods representing the complexity of these methods can be useful in raising awareness and in pushing researchers and practitioners to enhance and enrich their toolboxes to achieve a better and more profound knowledge of their field, so important in today's social and economic settings.

The book is organised as follows.

Chapter 1 discusses the static and dynamic attributes of a complex system, the basis for justifying the use of modelling and simulation techniques. When choosing a method for exploring a phenomenon or a subject, we inevitably make assumptions on their nature. These assumptions direct the way we formulate questions, or arrange theories and models, carry out empirical work and interpret evidence. It is well known that a complex system must be treated in a holistic way and that many of the conventional methods are unable to correctly provide insights on specific issues, problems and questions. This is also true in the study of tourism and other connected areas.

Chapter 2 contains a series of considerations on what modelling is and the importance and the usefulness of simulation activities. Models describe our beliefs about how the object of our studies functions, which are translated into some formal language. In some cases, this language is the language of mathematics, and the relationships we write may give full account of the peculiarities and the behaviours of our target. In most cases, however, we cannot easily produce relationships, or we are unable to solve the equations and we need to resort to numerical models. In either case, building a reliable model and running a successful simulation requires attention to a number of elements that are discussed in the chapter.

Chapter 3 describes the features and introduces the most used modelling methods. Conceptual, statistical, machine learning, network analytic, system dynamic and agent-based models are briefly sketched and worked examples are proposed to better grasp the modelling and simulation methods of interest to the tourism community. These are, rather obviously, all numerical computational methods that analyse several aspects, from the structural characteristics of a tourism destination, the most important object of study in the tourism domain, to the examination of different facets of the preferences, the needs and the behaviours of travellers and tourists.

Chapter 4 briefly outlines some of the most interesting advanced methods in the fields of network analysis and artificial intelligence. The most elementary methods used for analysing networks, although proven quite effective in the tourism domain, hardly scratch the surface of the issues in a complex environment like tourism. Multiple types of relationships and different temporal activations call for approaches that are able to render this multiplicity, and with so diverse systems a consistent way to assess the results is needed. Finally, the recent incredible results achieved in the field of artificial intelligence and machine learning are already hitting the field and it is important to have a basic understanding of their functioning and applicability.

Chapter 5 is dedicated to the intricate issue of choosing a modelling technique. Here, we describe some of the possible criteria that can be used in selecting one or more of the different methods and how to combine them into a rational and effective approach. We also take a look at how developments and implementations can further assist in the difficult art of modelling.

Chapter 6 contains case studies that show how different approaches can be combined to create models or simulations used to solve issues or gather insights that are interesting from a theoretical and practical point of view.

The book ends with an appendix containing some further reading suggestions and an appendix with references to some of the most popular and used software programs, and some tutorials dedicated to those who want to start using the main techniques discussed in this work.

1 Systems and Tourism Systems

Introduction

In this initial chapter, we set out the conceptual bases for the domain of modelling and simulations. Understanding how systems are structured and the relationships between their elements, and the possibility to exploit this understanding is the essence of scientific method. Study outcomes have allowed us to better realise how many phenomena have progressed and improved our abilities (even if sometimes limited) to predict future patterns. These enquiries have also allowed us to find similarities in different situations, thereby extending our ability to describe events and solve problems. Probably more than any other human activity, tourism is a complex phenomenon composed of different entities (companies, groups, individuals, etc.) and resources interacting in non-trivial ways to satisfy the needs and wishes of its users, the tourists.

Managing and governing this phenomenon and the systems that are part of it requires approaches that differ from traditional perspectives, and attempts need to be made at experimenting for possible solutions or arrangements that cannot be obtained by 'experimenting' in real life. Thus, the need for a set of 'artificial' ways to achieve these objectives has led to the building of numeric and computerised models that form the basis for simulating different settings. These, in turn, pose the necessity for a systemic holistic view, which is more suitable than traditional reductionist approaches. This perspective is rooted in the research tradition of what is today known as *complexity science* (Bertuglia & Vaio, 2005; Edmonds, 2000).

The main object of our studies is a system: 'an interconnected set of elements that is coherently organized in a way that achieves something' (Meadows, 2008: 11). This definition accurately describes the three important components – the elements, the interconnections between them and the function or purpose of a system.

In the tourism and hospitality domains, the set of interconnected actors, from travel agencies to accommodation entities, from services such as restaurants and private tours to transportation, and public actors as well as associations and the local population form a system with the objective to promote or favour tourism in a specific destination to increase the social and economic wealth of the entities involved. If we think of tourism as a system, then the behaviour of the overall system depends on how all its parts (accommodation, services, transportation, etc.) interact and exchange information, financial flow, knowledge and strategies.

As will become clear in the following pages, a system is more than just the sum of its parts. Usually it is a dynamic entity that may exhibit adaptive and evolutionary behaviour. Therefore, it is important to consider a system in its entirety, to shift the attention from the parts to the whole.

Towards a Systemic View: A Short History

In the course of its history, our civilisation has set up and refined a relatively standard way of studying a phenomenon, tackling an issue or solving a problem. However, in many cases, the standard way is modified by individual convictions and viewpoints that, even if seldom defined completely or coherently, may have wide effects. Moreover, personal philosophical and epistemological beliefs have always played a crucial role in the history of science, and in many cases have greatly influenced the development of ideas and knowledge (Baggio, 2013).

The general approach is composed of a series of phases: (i) define an objective and the object of study; (ii) decide whether the knowledge and methods are sufficient to address it; (iii) explore what and how others have produced in comparable cases; (iv) plan and collect empirical evidence; (v) derive the appropriate conclusions; and, finally, (vi) outline an action to meet the aims of the work conducted. In doing so, researchers use a vast array of specific techniques, epistemological positions and philosophical beliefs (Losee, 2001).

In this scenario, one element has historically been well supported and accepted. When facing a big problem, a large system or a convoluted phenomenon, the best line of attack is to split it into smaller parts that can be more easily managed. Once all the partial results have been obtained, they can be recomposed to find a general solution. This notion is known as *reductionism*. It can be summarised in the words of René Descartes (1637) who formalised the idea. In *Discourse on Method* (1637: part II), he states that it is necessary 'to divide each of the difficulties under examination into as many parts as possible, and as might be necessary for its adequate solution'.

Moreover, in the *Regulae ad directionem ingenii* (rules for the direction of the mind), Descartes (1701) clearly states in rule V that 'Method consists entirely in the order and disposition of the objects towards which our mental vision must be directed if we would find out any truth. We shall comply with it exactly if we reduce involved and obscure propositions step by step to those that are simpler, and then starting with the intuitive apprehension of all those that are absolutely simple, attempt to ascend to the knowledge of all others by precisely similar steps'; and in rule XIII that 'If we are to understand a problem perfectly, we must free it from any superfluous conceptions, reduce it to its simplest terms, and by process of enumeration, split it up into its smallest possible parts'.

Reductionism, however, has a much longer history. It is rooted in ideas and concepts that evolved from the pre-Socratic attempts to find the universal principles that would explain nature and the quest for the ultimate constituents of matter. The whole Western tradition then elaborated on these concepts that were admirably distilled in the 16th and 17th centuries. Copernicus, Galileo, Descartes, Bacon and Kepler came to a rigorous formulation of the method needed to give a truthful meaning to *science*. This work was very effectively refined by Isaac Newton (1687) in his *Philosophiae Naturalis Principia Mathematica*. The book was so successful and so widely distributed that scholars of all disciplines started to apply the same ideas to their own field of study, especially in those domains that did not have a strong empirical tradition such as the study of human societies and actions.

The reasons for this wide influence were, essentially, the coherence and apparent completeness of the Newtonian proposal coupled with its agreement with intuition and common sense. In the following years, many scientists tried to extend this perspective to other environments. Scholars such as Thomas Hobbes, David Hume, Adolphe Quetelet and Auguste Comte worked with the objective to explain aggregate human behaviour using analogies from the world of physics, and employing its laws. Vilfredo Pareto and Adam Smith, for example, presented and agreed on a 'utilitarian' view of the social world, which suggests that the use-value of any good can be fully reflected in its exchange-value (price). With these bases, they tried to adapt the mechanical paradigm to the field of economics. The idea of defining universal laws, setting mathematical analytic expressions and formulating gravity models or terms such as equilibrium is directly derived from the *Principia*.

The universality of Newton's proposals, however, was questioned when the scientific community began to realise that going beyond simple individual objects created a number of additional variables due to mutual interactions, so that solutions to even simple-looking problems could not be easily obtained unless the 'finer details' in the mathematical formulation were disregarded and descriptions limited to a simplified and linearised description.

For example, the gravitational theory was accurate in dealing with simple sets of objects, but failed when applied to more numerous assemblies. The motion of planets in the solar system was initially well described, but some deviations, such as the curious perturbations observed in Mercury's orbit, could not find a proper place in the model. Deeper investigations showed that an increase in the number of bodies in a gravitational system, made the motion of the different elements almost unpredictable. Poincaré (1884) eventually realised that even a small three-body system can produce complicated outcomes and that the equations describing it become extremely difficult to determine and practically unsolvable. The stability conditions for equilibrium in the motion of a system were later studied and characterised by Lyapunov (1892). He provided the first evidence of the fact that, in some cases, even minor changes in initial conditions, described by deterministic relationships, can result in widely differing system evolutionary trajectories. This is what we term *dynamical instability* or *sensitivity* to initial conditions, and today we identify as *chaos*.

The issue of dealing with a system composed of many elements gained attention in the first half of the 19th century. The practical issue of increasing the efficiency of the newly developed steam engine led several scientists to leave the path drawn by Newton and approach the matter from a different point of view. The problem was to study the behaviour of a gas in which a very large number of particles interact (1 L of air contains about $3 \cdot 10^{22}$ molecules). Explicitly writing numerous equations and solving the problem was impossible. Thus, statistical techniques were thought to be the only possible way to tackle the problem. Nicolas Sadi Carnot, James Joule, Rudolf Clausius and William Thomson (Lord Kelvin) created the new discipline, thermodynamics, based on these ideas.

Their results were successful and, elaborating on them at the end of the 19th century, James Clerk Maxwell, Ludwig Boltzmann and Josiah Willard Gibbs structured the matter into what is known today as *statistical mechanics* (or statistical physics). The central idea is that the knowledge of an incomplete set of measurements of some system's properties can be used to find the probability distributions for other properties of the system. For example, knowing the number of molecules in a gas in a certain volume and its temperature at thermal equilibrium (i.e. when no spatial or temporal variations in temperature exist), it is possible to calculate the pressure, the specific heat and other quantities.

Statistical physics is a very rigorous formal framework for studying the properties of many-body systems (i.e. composed of many interacting particles), where macroscopic properties are statistically derived from extensive (dependent on the size or the amount of material) and intensive (independent of the amount of material) quantities, and their microscopic properties can be described in terms of probability distributions. Furthermore, it is possible to have a better understanding of the conditions in which critical modifications of a system or sudden changes to its state (phase transitions) occur. This understanding of phase transitions and critical phenomena led to the development of two important new concepts: *universality* and *scaling* (Amaral & Ottino, 2004).

When studying critical phenomena, or critical conditions in a system's evolution, a set of relations, called *scaling laws*, may be determined to help relate the various critical-point features by portraying the behaviour of some system parameters and of the response functions. The predictions of a scaling hypothesis are supported by a wide range of experimental work, and by numerous calculations on model systems (Kadanoff, 1990).

The concept of universality has the basic objective of capturing the essence of different systems and classifying them into distinct classes. The universality of critical behaviour drives explorations to consider the features of microscopic relationships as important in determining critical-point exponents and scaling functions. Statistical approaches can thus be very effective in systems when the number of degrees of freedom (and elements described by several variables) is so large that an exact solution is neither practical nor possible. Even in cases where it is possible to use analytical approximations, most current research utilises the processing power of modern computers to simulate numerical solutions. Here, too, experimental work and numerical simulations have thoroughly supported the idea (Stanley, 1999).

The main result and the power of this approach are in recognition that many systems exhibit universal properties that are independent of the specific form of their constituents. This suggests the hypothesis that certain universal laws may apply to many different types of systems, be they social, economic, natural or artificial. For example, biotic environments can be well described in terms of their food webs. Analyses of a wide number of such systems, examined in terms of their networks composed of different species and their predation relationships, show remarkable similarities in the shapes (topologies) of these networks (Garlaschelli *et al.*, 2003). This happens independently because of *apparently* signifi-cant differences in factors such as size, hierarchical organisation, specific environments and history. The universality and scaling hypotheses thus seem valid in this field and might open the way to reconsidering the pos-sibility of establishing some general treatment for the problems in envi-ronmental engineering.

In other words, these assumptions give us the basis to justify an exten-sive use of analogy, so that an inference can be drawn based on a similar-ity in certain characteristics of different systems, typically their structural configuration (topology). That is to say, if a system or a process A is known to have certain traits, and if a system or process B is known to have at least some of those traits, an inference is drawn that B also has the others. Often, mathematical models can be built and numerical simula-tions run to transfer information from one particular system to another particular system (Daniel, 1955; Gentner, 2002).

Complex Adaptive Systems

A systemic view is focused on considering a configuration of elements joined together by a web of relationships, sensitive to external forces that may modify its structure or behaviour. In this approach, we leave the tradi-tional idea of cause and effect, directly connected with that of predictability, and use statistical methods for creating possible scenarios and assign them a probability to happen. This is the idea of *complex adaptive systems* (CAS).

The natural language concept of complexity has several meanings, usu-ally associated with the size and number of components in a system. There is still no universally accepted definition nor a rigorous theoretical formali-sation of complexity; however, intuitively, we may characterise a complex system as 'a system for which it is difficult, if not impossible to reduce the number of parameters or characterizing variables without losing its essen-tial global functional properties' (Pavard & Dugdale, 2000: 40).

Basically, we consider that a system is complex if its parts interact in a non-linear manner. Rarely are there simple cause-and-effect relationships between elements, and a small spur may cause a large effect or no effect at all. The non-linearity of interactions generates a series of specific proper-ties that characterise the complexity of a system's behaviour.

Broadly speaking, systems can be categorised into simple, compli-cated and complex systems. The most important difference between these categories is that simple and complicated systems are predictable and results are repeatable: that is, if a specific strategy, assemblage, policy, etc., initially worked, it will always work for the same system thereafter.

Complex systems, on the other hand, are neither fully predictable nor fully repeatable. Further, simple and complicated systems can be understood by assessing how each part that composes the system works and by analysing the details. However, when dealing with complex systems, we can only achieve a partial understanding of the system, and only by looking at the behaviour of the system in its entirety. That is, we would not be able to understand the interactions and outcomes of a tourism destination by reducing it to specific entities and analysing them, but only by observing or analysing how the overall system behaves. Even then, we would only achieve a partial understanding of the overall system.

Complicated systems are systems in which there are clear and defined cause–effect relationships, but a system may have many components and, as such, there may be multiple ways to achieve a solution to a problem. However, there is only one correct solution. In other words, complicated systems are the sum of multiple simple systems, and thus we can understand and achieve a solution by decomposing the system into multiple simple systems, solving problems that have only one solution and then putting the simple system back together. This shows the importance of coordinating how different parts should be assembled, implying a further step than just putting simple systems back together. In a complex system, multiple possibilities exist that have many interacting parts, are dynamic and adapt themselves to interacting with local conditions, often producing new (emergent) structures and behaviours. As Stacey (1996: 10) states, in a complex adaptive system – the term used to denote this type of system – the parts 'interact with each other according to sets of rules that require them to examine and respond to each other's behaviour in order to improve their behaviour and thus the behaviour of the system they comprise'. A schematic view of the main differences between simple, complicated and complex systems is shown in Table 1.1.

Table 1.1 Characteristics of simple, complicated and complex systems

Feature	Simple	Complicated	Complex
Number of elements	Few	Many	Moderate to many
Similarity of elements	Identical (or of same type)	Partly different	Partly or entirely different
Variability over time	No	No	Yes
Number of relationships	Few	Moderate to many	Moderate to many
Type of relationships	Linear	Linear	Non-linear
Example	Pendulum, wheel	Car, airplane	Ecosystem, tourism destination

Complex systems are difficult to define and there is no consensus on a possible formal characterisation. However, scholars and practitioners in the field have provided a long list of complex systems characteristics, apart from size considerations. The most relevant are (Bar-Yam, 1997; Waldrop, 1992):

- *Distributed nature*: Properties and functions cannot be precisely located and there may be redundancies and overlaps.
- *Non-determinism*: No precise anticipation of behaviour is possible, even knowing the functional relationships between elements. The dependence of the system's behaviour from the initial conditions is extremely sensitive, and the only predictions that can be made are probabilistic.
- *Presence of feedback cycles (positive or negative)*: Relationships among elements become more important than their own specific characteristics and reinforcing or balancing actions can influence the overall behaviour of the system.
- *Emergence and self-organisation*: Several properties are not directly accessible (identifiable or predictable) from knowledge of the components. Global or local structures may emerge when certain parameters (a system's characteristics) pass a critical threshold. The system is capable of absorbing the shock and remaining in a given state, regaining its original state or adapting to new conditions unpredictably fast (system is resilient). From an empirical point of view, it is virtually impossible to determine why a system prefers one specific configuration over possible alternatives, or the type of perturbations that may disrupt or be absorbed.
- *Self-similarity*: The system considered will look like itself on a different scale, if magnified or made smaller in a suitable way. The self-similarity is evidence of the possible internal complex dynamics of a system.
- *Limited decomposability*: The properties of a dynamic structure cannot be studied by decomposing it into functionally stable parts. Its interaction with the environment and its properties of self-organisation allow it to functionally restructure itself; only a 'whole system' approach can explain characteristics and behaviours.

In short, following Cilliers (1998), it is possible to characterise a system as complex and adaptive if the system has the following main properties (see Figure 1.1):

- many elements form the system;
- interactions among the elements are non-linear and usually have a short range;
- there are loops in the interactions;
- complex systems are usually open and their state is far from equilibrium;
- complex systems have a history, their 'future' behaviour depends on their past behaviour;
- each element is unaware of the behaviour of the system as a whole, reacting only to locally available information.

Examples of CAS include many real-world ensembles: the patterns of birds in flight or the interactions of various life forms in an ecosystem, the behaviour of consumers in a retail environment, people and groups in a community, economic exchange processes, the stock market, the weather, earthquakes, traffic jams, the immune system, river networks, zebra stripes, seashell patterns and many others.

A CAS is a dynamical system and is, therefore, subject to evolutionary modifications that may be characterised by two variables: an order parameter and a control parameter. The first parameter represents the internal structure of the system, capturing its intrinsic order. The second parameter is an external variable that can be used to induce transitions in a system. For example, let us consider a certain volume of water close to boiling point. By increasing the temperature (providing energy [heat]

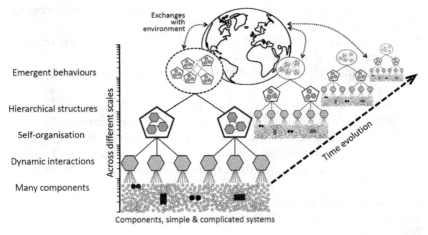

Figure 1.1 A graphical rendering of the main concepts related to complex systems

to the system), it is possible to bring the water to boiling point. At the critical temperature Tc = 100°C, the water starts boiling and the order parameter undergoes an abrupt change. It has a zero value in the random state (above the transition temperature) and takes on a nonzero value in the ordered state (below the transition). More generally, a variation of the order parameter can lead the system to a critical point (bifurcation) beyond which several stable states may exist. The state will depend on small random fluctuations, amplified by a positive feedback. It is impossible to determine or to control which state will be attained in a specific empirical system. Not even the control parameter (by itself) can be used to predict the system dynamics. Nonetheless, it is possible to sketch a general dependency of the *global conditions* of a system on a control parameter.

Starting from a stability condition, a system will evolve when the control parameter changes. The system passes through a periodic state, to a situation in which it exhibits complex behaviour, to a completely chaotic state. A simple example is that of a faucet. We can consider it a system controlled by the flow of water (the control parameter). Simple fluid dynamics tells us that we can predict flow patterns in different situations by looking at a quantity called Reynolds number R (roughly dependent on the quantity of fluid passing through the pipe). At low R, the resulting flow has a steady laminar behaviour. Increasing the quantity of water, we enter a stage in which the behaviour is 'transitional', some limited turbulence appears. By increasing the quantity of water further, we arrive at a stage of high turbulence, a chaotic state in which the fluid changes speed and direction unpredictably. One more example can be explored using a logistic map. This is the equation that governs the growth of a population before attenuating as it reaches its carrying capacity. The equation has as its parameter the growth rate r, the population level at any given time is a function of the growth rate and the previous population. Depending on the value of this parameter, the population may become extinct (r too low) or it may stay stable, fluctuating across two or more values (a complex phase, at the *edge of chaos*) or it may rapidly change across a series of expansions and reductions (a chaotic phase, Figure 1.2).

Many of the real systems we know live at the boundary between complexity and chaos. This situation is frequently referred to as the *edge of chaos*, where a system is in a state of fragile equilibrium, on the threshold of collapsing into a rapidly changing state, which may set off a new dynamic phase (Waldrop, 1992).

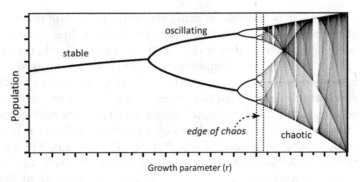

Figure 1.2 A logistic map

Complexity and wicked problems

The study of complex systems calls for a new strategy in which multiple methods stemming from multiple traditions and disciplines are employed, and where local and academic knowledge are integrated. Decision-making activities and how we make decisions are dependent on the type of system we are considering or are embedded in. That is, the problems we need to solve are strongly linked to the 'complexity' of the system.

Problems can be defined on a continuum from simple to complicated to complex or *wicked*, those that are difficult or impossible to solve because of incomplete, contradictory or changing requirements, or situations that are often difficult to recognise, or because they are embedded in complex systems (Alford & Head, 2017; Baggio *et al.*, 2019). More precisely, problems can be defined depending on a specific system's characteristics such as whether and to what extent the behaviour of the system is predictable, characterised by tipping points and non-linear dynamics, has more or less interacting elements and solutions that are more or less clear. At the same time, however, solving problems also depends on the amount of knowledge needed to solve them, and on whether that knowledge is fragmented across different stakeholder groups, or stakeholders have multiple or conflicting objectives. Finally, problems may form 'Russian dolls', that is, they can be nested in one another just like a set of nested wooden dolls.

Simple problems and complicated problems are those for which technical solutions are (more or less) easily available. If a car or a bus breaks down, a visit to a repair shop is the obvious answer, and if a hotel room's temperature is too high, a fix to the air-conditioner will solve the problem.

Simple, Complicated and Complex Problems		
Following a Recipe	Sending a Rocket to the Moon	Raising a Child
The recipe is essential	Formulae are critical and necessary	Formulae have a limited application
Recipes are tested to assure easy replication	Sending one rocket increases assurance that the next will be OK	Raising one child provides experience but no assurance of success with the next
No particular expertise is required. But cooking expertise increases success rate	High levels of expertise in a variety of fields are necessary for success	Expertise can contribute but is neither necessary nor sufficient to assure success
Recipes produce standardized products	Rockets are similar in critical ways	Every child is unique and must be understood as an individual
The best recipes give good results every time	There is a high degree of certainty of outcome	Uncertainty of outcome remains
Optimistic approach to problem possible	Optimistic approach to problem possible	Optimistic approach to problem possible

Figure 1.3 Problems and their characteristics (After Glouberman & Zimmerman, 2004)

More common, however, are those problems that are precisely defined, such as traffic congestion or pollution on beaches, but that do not have commonly agreed on and clear-cut solutions. Solving these issues involves not only the possibility of proposing a solution but also a continuous effort in refining such solutions based on previous outcomes. Wicked problems are problems that cannot be clearly defined nor have clear solutions and require continuous learning, often between different decision makers, to continuously refine the problem statement as well as the potential solutions. They typically require the collaboration and coordination of multiple types of competences (DeFries & Nagendra, 2017). To further clarify different types of problems, some well-known examples are provided in Figure 1.3 (Glouberman & Zimmerman, 2004: 22).

Tourism and Tourism Systems

As an economic activity, the tourism sector shares many of the characteristics we have identified as typical of *complexity*. Complexity is characterised by dynamically emergent behaviours generated by intricate relationships between many diverse components: public and private organisations, small and large companies, individuals and environmental resources. Another issue of complexity is that it is almost impossible to fully delimit an object of study because of the tourism sector's strong connections with the wider political, social and economic environment. The sector also contains a number of 'subsectors', including hospitality, leisure, entertainment and recreation, and specific tourism activities, such as cultural, medical, gastronomic, ecological and so on.

Economists maintain that tourism is not an 'industry', since it does not produce a product (they would say that there is no production function) and there are no consistently measurable outputs and no common structure or organisation across countries or even within the same country (see, e.g. OECD, 2000). Moreover, tourism activities straddle several traditional economic sectors and are generally not considered, as a whole, in the national accounts. According to the World Tourism Organisation's definition, tourism covers: 'the activities of persons travelling to and staying in places outside their usual environment for not more than one consecutive year for leisure, business and other purposes' (UNWTO, 1995). This definition, however, looks ambiguous if examined through the glasses of a scientist. Too many different elements and interpretations are contained in a possible definition of tourism. This is the main reason why all interested parties responsible for *measuring* this phenomenon are challenged when it comes to distinguishing the 'units' to be accounted for, and the lack of a formal definition poses many problems to those seeking to understand tourism's basic functioning and to predict future behaviours.

Even if not always explicitly defined as such, the idea that tourism is a CAS has circulated for a long time (Leiper, 1979). Despite the lack of a clear and rigorous characterisation, many researchers and practitioners have devised several models, methods and approaches with the objective to help understand structures and dynamic evolutions, and have provided the means to manage systems, to predict their effects and to optimise their functions. Many authors have employed complexity science approaches to tourism, starting from the recognition of the *complexity* and instability of these systems that may lead to chaotic behaviours (Edgar & Nisbet, 1996; Faulkner & Russell, 1997; McKercher, 1999). Therefore, it has been claimed that the holistic approach is the only reasonable approach to studying the phenomenon and its components. As Farrell and Twining-Ward (2004: 287) state in their seminal paper: 'Both social and natural systems are considered complex and real entities, tightly integrated, and functioning together as nonlinear, evolving socioecological systems. To move tourism towards such a transition, researchers, consultants, managers, and stakeholders need to understand complex systems through integrative and nonlinear approaches; otherwise progress will be hampered, and results distorted, incomplete and devoid of full meaning'.

The most important component in the tourism domain is the destination. The concept of a tourism destination implies a systemic attitude in tourism analyses, an approach in which the focus is on the activities and the strategies to foster the development of an area delineated as a system of actors combining forces to supply integrated tourism products and services.

The challenge of defining what a destination is has led to many definitions over time (Pearce, 2014). The UNWTO (2002) Working Group on Destination Management uses the following definition: 'a local tourism destination is a physical space in which a visitor spends at least one overnight. It includes tourism products such as support services and attractions, and tourism resources within one day's return travel time. It has physical and administrative boundaries defining its management, and images and perceptions defining its market competitiveness. Local destinations incorporate various stakeholders often including a host community and can nest and network to form larger destinations'.

This definition addresses many of the foregoing issues. Many different elements are contained in this definition that characterises the basic elements of a tourism system (companies, products, tourists). The effort to overcome a purely geographical attribute is evident, as is the attempt to recognise the contributing factors, not only businesses, but also the local community and a managing entity. These conform well to the idea of a district envisaged by Becattini (1990) and his followers.

In essence, a destination is a complex networked system, where local actors (public and private) and organisations work together with the objective of satisfying the needs and wishes of those who visit the place. Destination clusters generally establish spontaneously and evolve and change over time, driven by both internal and external factors. They are not isolated entities, but open systems with complex links to several similar or diverse structures. New competitive products and services are frequently developed in cooperation with other ensembles, and the interface between different agglomerations allows the creation of new value (Nordin, 2003).

The complex nature of this agglomeration of entities and the difficulty of stating a geographical or sectoral delimitation make the analysis quite difficult. On the other hand, the local population and the local social system, the various institutional entities (local or country governments, associations, etc.) and other organisations play a role, although not directly of a *touristic* nature, which is essential for the good functioning of the system as a tourism destination. Moreover, today's advances in information and communication technologies have eased several constraints in terms of space and time so that virtual groupings with entities external to the specific area can be established, thus overcoming the need for strict physical proximity, leading to the necessity to widen the perspective and consider these external entities as digital ecosystems (Boes *et al.*, 2016; Buhalis & Amaranggana, 2014).

Managing such an ecosystem means finding a way to 'direct' a complex system that, almost by definition, is relatively unmanageable. Essentially, it corresponds to finding a control structure that is able to produce order by inducing certain relations among its elements with the effect of reducing the indistinguishability of the states of the system and augment its stability. On the other hand, seeking a too stable equilibrium is considered detrimental to the development of such structures; evolution and growth can only be possible in conditions that are at the boundary between order and chaos (Rosenhead, 1998: 3.1): 'Rather than trying to consolidate stable equilibrium, the organisation should aim to position itself in a region of bounded instability, to seek the edge of chaos. The organisation should welcome disorder as a partner, use instability positively. In this way new possible futures for the organisation will emerge, arising out of the (controlled) ferment of ideas which it should try to provoke. Instead of a perfectly planned corporate death, the released creativity leads to an organisation which continuously re-invents itself. Members of an organisation in equilibrium with its environment are locked into stable work patterns and attitudes; far from equilibrium, behaviour can be changed more easily'.

Not being in equilibrium is also akin to being loosely coupled, or in better terms, avoiding rigidity and locking in positions that may render future change impossible – think of the difference between bamboo and a large oak tree. The bamboo may constantly bend and change position even under very low-intensity perturbations (i.e. wind) but it is able to withstand higher-level perturbance (wind) intensity and frequency. On the other hand, an oak tree may be in 'equilibrium' or static for higher levels of wind but once the perturbance reaches a certain intensity, the oak tree will fall down. A loose coupling within organisations and an ability to adapt is key to their success. In fact, as stated by Levinthal (1997), 'the degree to which a firm is composed of loosely coupled subsystems provides important insights in understanding the variation in survival among firms facing a changing environment. Tightly coupled organizations can not engage in exploration without foregoing the benefits of exploitation. For a tightly coupled organization, efforts at search and experimentation tend to negate the advantages and wisdom associated with established policies and thereby place the organization at risk of failure. In contrast, more loosely coupled organizations can exploit the fruits of past wisdom while exploiting alternative bases of future viability'.

Nevertheless, the possibility to govern a complex system still exists. Good knowledge of a system and its main dynamic behaviours allow its controllability, that is the possibility not only to direct, but also to explore how to steer it to a desired final state or trajectory or a desired collective behaviour. To accomplish this, a good modelling practice is crucial as it not only allows a basic understanding but also presents the bases for simulating the effects of possible actions and building scenarios that can help make informed decisions.

Different leadership styles are important for making and managing decisions within complex systems. In fact, following Glouberman and Zimmerman (2004), one can discern that in complex systems, leadership needs to facilitate relationship building, increasing the ability of teams to collectively make sense of the interactions that are of interest to the problem at hand, encouraging learning practices, as well as a less strict top-down approach, thereby allowing communities of practice to experiment with novel ideas. Finally, detecting small changes in a system and understanding emergent directions can be crucial in avoiding possible disruptions.

Concluding Remarks

In a dynamic and rapidly changing world, understanding phenomena and the systems governing them and attempting to predict their evolution has become a daunting task. This is particularly true for the tourism domain. A largely multifaceted, inhomogeneous and poorly defined domain, the intricacy of the relationships between the different parts and with the external environment generates configurations and effects that often seem largely unpredictable.

For many years (and partially still today), researchers and practitioners have approached the issue using a reductionist approach and have considered that such complexity could be understood by looking at its principal components. Moreover, relying on these assumptions, they have designed a wealth of techniques for the task.

In this initial chapter, we have discussed what a system is and what are the main features of complex adaptive systems that form the basis for the following chapters. The idea is that complex systems, such as those in the tourism domain, need accurate models that are able to explain their main characteristics and form the basis for a deeper understanding. These also play a crucial role in the practice of simulating possible changes, variations or adaptations and building scenarios that can better inform decisions and plans with the objective of a balanced development of a tourism system.

In fact, besides its obvious theoretical appeal, this approach has already demonstrated good *practical* value and continues to do so. From this perspective, the toolbox available today is crammed. Several techniques have been developed to deal with the task of studying the structural and dynamic features of a complex system. The starting point is the formulation of a suitable model.

References

Alford, J. and Head, B.W. (2017) Wicked and less wicked problems: A typology and a contingency framework. *Policy and Society* 36 (3), 397–413.

Amaral, L.A.N. and Ottino, J.M. (2004) Complex networks: Augmenting the framework for the study of complex systems. *The European Physical Journal B* 38, 147–162.

Baggio, J.A., Freeman, J., Coyle, T.R., Nguyen, T.T., Hancock, D., Elpers, K.E., Nabity, S., Dengah Ii, H.J.F. and Pillow, D. (2019) The importance of cognitive diversity for sustaining the commons. *Nature Communications* 10 (1), art. 875.

Baggio, R. (2013) Oriental and occidental approaches to complex tourism systems. *Tourism Planning and Development* 10 (2), 217–227.

Bar-Yam, Y. (1997) *Dynamics of Complex Systems*. Reading, MA: Addison-Wesley.

Becattini, G. (1990) The Marshallian industrial district as a socio-economic notion. In F. Pyke, G. Becattini and W. Sengenberger (eds) *Industrial Districts and Inter-firm Co-operation in Italy* (pp. 37–51). Geneva: International Institute for Labour Studies.

Bertuglia, C.S. and Vaio, F. (2005) *Nonlinearity, Chaos, and Complexity: The Dynamics of Natural and Social Systems*. Oxford: Oxford University Press.

Boes, K., Buhalis, D. and Inversini, A. (2016) Smart tourism destinations: Ecosystems for tourism destination competitiveness. *International Journal of Tourism Cities* 2 (2), 108–124.

Buhalis, D. and Amaranggana, A. (2014) Smart tourism destinations. In P. Xiang and I. Tussyadiah (eds) *Information and Communication Technologies in Tourism 2014 (Proceedings of the International Conference in Dublin, Ireland, January 21–24)* (pp. 553–564). Berlin-Heidelberg: Springer.

Cilliers, P. (1998) *Complexity and Postmodernism: Understanding Complex Systems*. London: Routledge.

Daniel, V. (1955) The uses and abuses of analogy. *Operations Research Quarterly* 6 (1), 32–46.

DeFries, R. and Nagendra, H. (2017) Ecosystem management as a wicked problem. *Science* 356 (6335), 265–270.

Descartes, R. (1637) *Discours de la méthode pour bien conduire sa raison, et chercher la verité dans les sciences*. Leiden: De l'imprimerie de Ian Maire.

Descartes, R. (1701) *Regulae ad directionem ingenii*. In *Des-Cartes Opuscula posthuma, physica & mathematica*. Amsterdam: P. and J. Blaeu.

Edgar, D.A. and Nisbet, L. (1996) A matter of chaos: Some issues for hospitality businesses. *International Journal of Contemporary Hospitality Management* 8 (2), 6–9.

Edmonds, B. (2000) Complexity and scientific modelling. *Foundations of Science* 5, 379–390.

Farrell, B.H. and Twining-Ward, L. (2004) Reconceptualizing tourism. *Annals of Tourism Research* 31 (2), 274–295.

Faulkner, B. and Russell, R. (1997) Chaos and complexity in tourism: In search of a new perspective. *Pacific Tourism Review* 1, 93–102.

Garlaschelli, D., Caldarelli, G. and Pietronero, L. (2003) Universal scaling relations in food webs. *Nature* 423 (6936), 165–168.

Gentner, D. (2002) Analogy in scientific discovery: The case of Johannes Kepler. In L. Magnani and N.J. Nersessian (eds) *Model-Based Reasoning: Science, Technology, Values* (pp. 21–39). New York: Kluwer Academic/Plenum.

Glouberman, S. and Zimmerman, B. (2004) Complicated and complex systems: What would successful reform of Medicare look like? In P.-G. Forest, G.P. Marchildon and T. McIntosh (eds) *Changing Health Care in Canada: Romanow Papers, Vol. 2* (pp. 21–53). Toronto: University of Toronto Press.

Kadanoff, K.P. (1990) Scaling and universality in statistical physics. *Physica A* 163 (1), 1–14.

Leiper, N. (1979) The framework of tourism: Towards a definition of tourism, tourist, and the tourist industry. *Annals of Tourism Research* 6 (4), 390–407.

Levinthal, D.A. (1997) Adaptation on rugged landscapes. *Management Science* 43, 934–950.

Losee, J. (2001) *A Historical Introduction to the Philosophy of Science* (4th edn). Oxford: Oxford University Press.

Lyapunov, A.M. (1892) *General Problem on Motion Stability* (Vol. 11). Kharkov: Kharkovskoye Matematicheskoe Obshchestvo. [In Russian.]

McKercher, B. (1999) A chaos approach to tourism. *Tourism Management* 20, 425–434.

Meadows, D.H. (2008) *Thinking in Systems: A Primer*. White River Junction, VT: Chelsea Green.

Newton, I. (1687) *Philosophiae Naturalis Principia Mathematica*. London: jussi Societatus Regiae ac typis Josephi Streater.

Nordin, S. (2003) *Tourism Clustering and Innovations* (Scientific Report No. 2003:14). Ostersund: ETOUR – The European Tourism Research Institute. See http://www.diva-portal.org/smash/get/diva2:352389/FULLTEXT01.pdf (accessed November 2006).

OECD (2000) *Measuring the Role of Tourism in OECD Economies*. Paris: OECD.

Pavard, B. and Dugdale, J. (2000) The contribution of complexity theory to the study of socio-technical cooperative systems. Third International Conference on Complex Systems, Nashua, NH, 21–26 May. See http://membres-lig.imag.fr/dugdale/papers/Contribution%20of%20complexity%20theory1.pdf (accessed October 2005).

Pearce, D.G. (2014) Toward an integrative conceptual framework of destinations. *Journal of Travel Research* 53 (2), 141–153.

Poincaré, H. (1884) Sur certaines solutions particulières du problème des trois corps. *Bulletin Astronomique* 1, 63–74.

Rosenhead, J. (1998) *Complexity Theory and Management Practice. Science as Culture*. See https://sites.google.com/site/hetnieuwedenken/complexiteit (accessed December 2018).

Stacey, R.D. (1996) *Complexity and Creativity in Organizations*. San Francisco, CA: Berrett-Koehler.

Stanley, H.E. (1999) Scaling, universality, and renormalization: Three pillars of modern critical phenomena. *Reviews of Modern Physics* 71 (2), S358–S366.

UNWTO (1995) *Concepts, Definitions, and Classifications for Tourism Statistics. Technical Manual No. 1*. Madrid: World Tourism Organization.

UNWTO (2002) Terminology within destination management and quality. See http://marketintelligence.unwto.org/content/conceptual-framework-0 (accessed November 2005).

Waldrop, M. (1992) *Complexity: The Emerging Science and the Edge of Order and Chaos*. London: Simon & Schuster.

2 Models and Modelling

Introduction

What is a model and why should we model are important questions when it comes to understanding and assessing how specific decisions may affect future outcomes, especially within the context of complex systems. Modelling comes from the Latin word *modellus* and its original meaning was 'standard' or small measure. However, in the 1630s, the word *modelle* and then *model* started to be used to describe a 'standard for imitation or comparison'. Thus, models are a way to represent reality. The ability to abstract and represent reality is probably one of the most important features of our species (*Homo sapiens*), and representations of the real world have been around since the Stone Age. Drawings discovered in caves can be thought of as 'models' of specific activities. They capture specific properties (human, spear, animal) in order to represent a more complex real-world situation (the hunt). In this sense, drawings, words and numbers are models as they are a simplified representation of reality and help us make sense and explain reality.

Historically, one of the first and most important models, considering its effect on generations of scientists, can be dated back to Ptolemy, possibly inspired by the description of celestial objects as circles suggested by Pythagoras. We can ascribe to Ptolemy one of the first mathematical models of the solar system, where he employed the use of a base circle (deferent) and epicycles to describe and predict the movement of the planets as well as the sun and the moon (Figure 2.1).

This simple model's ability to predict the movement of planets was accurate for the time. The theoretical model also had an important 'practical' application in the development of an instrument, the astrolabe, which played a crucial role in the toolset of medieval astronomers and navigators from the 12th to the 17th centuries (Figure 2.2). A two-dimensional representation of the celestial sphere, which in addition to representing the sky

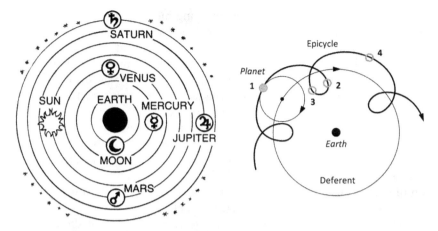

Figure 2.1 The Ptolemaic model of the solar system (left, after Foresman, 2008) and the deferent/epicycle model for the motion of a planet

and its most important bodies, allowed the measurement and calculation of the position of celestial objects, determined the time of observation, predicted events such as equinoxes and solstices, converted coordinates between different systems and many other. It can be thought of as one of the first (analogue) computers. It was over 1500 years before Copernicus and Kepler proposed and implemented a more refined and conceptually simpler model to describe and predict the motion of celestial objects.

Figure 2.2 Drawing of a medieval planispheric astrolabe (after Skeat, 2014)

Representing reality, or better, building models to represent real-life phenomena and address related issues, has been key in the development of human societies and our knowledge of the world. In fact, the ability to model, especially based on rules and mathematical equations, is highlighted by the fact that in multiple regions of the ancient world, methods were devised almost simultaneously to represent the universe in a simplified way (e.g. China, India and Persia). Generally speaking, a model always starts with an actual phenomenon, object or problem. This starting point is then simplified to structure or clearly characterise causal (or co-evolutionary) pathways using only what are considered the fundamental components of what is to be characterised or represented. As a simplified version of reality, models thus have limited usefulness defined by the specific scope of the model (or model purpose).

Models are used for different functions. In the case of Ptolemy and Kepler, models were used not only to explain, but also to predict specific outcomes. Explanation is often the objective of simple economic models (e.g. those dealing with the relations between demand and supply) as well as more complex models that try to make sense of dissension and cooperation in human societies, or predator–prey models that allow us to describe and clarify the interaction between species. Prediction, on the other hand, is the main objective of models related to weather patterns and extreme events such as hurricanes, as well as climate models that try to assess how increases in the concentration of greenhouse gas emissions affect temperature and climate.

Aside from explaining and predicting, models can be employed to make specific decisions. In this case, however, models need to explain and predict (at least probabilistically) before being able to be used effectively. The act of modelling itself forces one to be rigorous and precise in one's description of a phenomenon, thereby increasing the communicability of the phenomenon itself.

To this latter point, an often underappreciated value of models is their ability to communicate knowledge. Simplifying reality, for all intents and purposes, allows for easier communication of the real world or of events, even complex ones. According to some scholars (e.g. De Santillana & Von Dechend, 1977), ancient myths had this objective. Telling a nice story that included a simplified and comprehensible view of phenomena or situations (a model) was the main method of transferring precious and significant knowledge to those able to use it for speculative or practical purposes.

In a nutshell, given a specific problem, modelling is the act of abstracting the fundamental properties of a system and their relationship. Models are thus composed of elements, parameters (values that do not change during the time frame of the model) and variables (values that change due to feedback and interactions built within the model).

Modelling starts the process of thinking about and possibly solving (at least probabilistically) a problem and aids in solving problems that occur in the real world. Because modelling demands a clear problem definition, as well as an understanding of the interaction between a system's elements that may give rise to potential solutions, the act of modelling maps complex problems to a simplified representation, allowing careful study of the feedback and interdependencies that exist within the system to be modelled. Given that models are built on specific information, assumptions and issue, they are also a representation of what one perceives and believes about the functioning of the object of study. Such perceptions and beliefs are then translated into precise and clear language, whether formal (e.g. mathematical) or metaphorical (e.g. storytelling). Using a precise language adds to the modelling exercise, as it makes explicit the assumptions and rules used to describe the system, thereby allowing a better assessment not only of the solution to the proposed problem, but also of how the solution evolved.

Modelling, as a result, allows us to develop and increase our scientific understanding of a phenomena via a precise expression that permits an assessment of our current understanding about a system and may also highlight the information or knowledge required for a better evaluation of the situation that is being analysed. Through modelling, we are able to test how specific changes may affect the overall outcomes. That is, we can assess how a specific policy, or changes in climate, changes in infrastructures or specific decisions may affect the overall system. Doing this via modelling permits a clear cost reduction, but it is an essential practice when no experiments can be conducted or when there are ethical concerns about the implementation of such experiments. Modelling can greatly help decision makers and the decision-making process by making explicit the assumptions on which decisions are based and assessing the potential outcomes of those assumptions and decisions.

Generally speaking, models can be broadly divided into descriptive and analytical. Descriptive models rely on simulation and are often 'simulated'. Analytical models, in a pure definition, are made up of sets of equations describing the performance of a system. The results of the model functionally and clearly depend on the input of the model (the assumptions). Often, analytical models are used to understand and predict a system and have closed-form solutions. An important point is that in an analytical model, a given set of inputs will always give rise to the same outcome (results). In other words, analytical models are deterministic. However, when dealing with complex systems, often this type of model is not viable as equations cannot be written or if written cannot be solved, as stated by Poincaré (1884). In this case, we need to resort to a different description of the system and simulate how different input and relationship assumptions affect the outcome of interest. We need to turn to simulation where one models not only equations but also a set of rules that govern the system, and assess how these rules may or may not change due to their mutual interaction and the context in which the model has been conceived. For complex systems, simulations can often lead to a more realistic explanation of reality and improved predictions. Simulations allow for more complex experimentations, and the results are frequently represented by a 'distribution of results' rather than by single values. That is, in a simulation, results are probabilistic in nature. As previously stated, simulations and, more generally, descriptive models are an important way to facilitate our thinking and decision-making about the problem we want or need to tackle.

It is important to note that when we employ simulation and the results of our model are stochastic, it is almost always impossible to formulate specific, much less causal, statements such as policy X always leads to outcome Y. When employing simulations, we generate statements about the probability of a certain outcome given specific parameters. Hence, the results of simulations should be interpreted as 'on average, policy X leads to outcome Y' or 'in $N\%$ of the cases, policy X has led to outcome Y; however, in $M\%$ of the cases, the outcome was Z'. If we are to aid decision makers on whether to adopt specific policies, rules or incentives, we need to assess whether that policy, rule or incentive, for all possible outcomes, leads to a 'better' effect compared to alternatives or different conditions and under different parameters. If this happens, we have what is called a *dominant strategy*.

Within the realm of simulation models, different techniques exist that may or may not attract modelers depending on the problem to be solved. In fact, one can think of the appropriateness of a specific modelling technique depending on the level of aggregation required by the problem at hand (see Figure 2.3 where system dynamic and agent-based models (ABMs)[1] are compared, and consult Borshchev & Filippov [2004]). Different modelling techniques are more appropriate depending on the level of aggregation needed to represent a phenomenon, as well as whether we are interested in considering global properties or letting the global

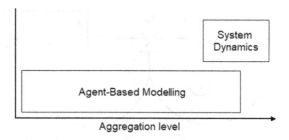

Figure 2.3 Level of aggregation and modelling technique

properties of the system emerge from individual interactions. More specifically, models can use different levels of information. The amount of information needed is determined by the level of aggregation needed to model a specific system or problem. Models can thus take into account all the interactions between parts mechanistically, as they consider all the mechanisms that influence specific changes or they can make allowances for aggregates assuming that changes at lower aggregation levels may (or may not) occur.

For example, if we think of a tourism destination, and we want to represent the adoption of new marketing strategies for accommodations with the objective of increasing occupancy, it is possible to analyse and study the feedback characteristics between accommodations that have adopted the new strategy, the effect of marketing the new strategy and the interactions between those who have and have not adopted the new strategy. In this case, we note or assume that different mechanisms leading to adoption may or may not exist, but we do not explicitly model them. Hence, we aggregate all the different accommodation preferences and adoption probabilities into one, and assume that all accommodations, at the aggregate level, will behave similarly. In other words, we are not interested in including the heterogeneity of individual accommodations or the processes that lead an accommodation to adopt or not the new strategy, but only the general diffusion of this new strategy. In this case, we would resort to model the system via system dynamics,[2] where entities are modelled as *stocks* and *flows*, roughly entities and dynamically changing variables. Figure 2.4 showcases an example of a stock and flow diagram representing a system dynamic model. Accommodations are positively influenced by a marketing campaign that increases the adoption rate. At the same time, accommodations adopting the new strategy influence others via word of mouth. The feedback between all accommodations and the adoption rate is negative (or balancing = B) as

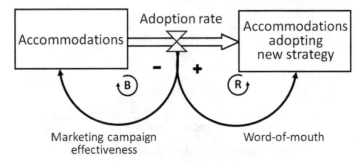

Figure 2.4 System dynamic representation of strategy adoption

the more accommodations that adopt, the less will adopt in the future (as there are less accommodations that have not adopted). On the other hand, the more accommodations that have adopted the new strategy, the more likely it is that other accommodations will adopt (reinforcing feedback = R). Note that in this case, we only model the general adoption process, and we do not examine individual accommodations preferences, capacities to adopt the new strategy or processes that may lead to that specific adoption (Figure 2.4).

On the other hand, if we are interested in the adoption of a new strategy to increase occupancy, and we are also interested in representing the heterogeneity in the decisions and the individual accommodations preference for adoption, we may resort to an ABM. Compared to system dynamics, agent-based modelling does not allow for the representation of a global system behaviour. Instead, the global system behaviour emerges from the individual interactions between the different actors, their preferences and how they react to the various inputs or stimuli. As Chapter 3 explains in more details, when employing an ABM, we model the behaviour of agents at the individual level (each actor may behave differently and interact differently with the environment and other actors). In this case, each entity is considered in relation to the specific process (a marketing campaign) as well as with other agents. Accommodations also interact with the environment in which they are embedded. The interaction between marketing, individual accommodations processes and the environment gives rise to the decision to adopt or not for each individual element. The environment represents the context of the accommodations. Agents, in general, can be defined depending on the problem at hand, thus an ABM can be used to represent multiple levels of aggregation (Figure 2.5).

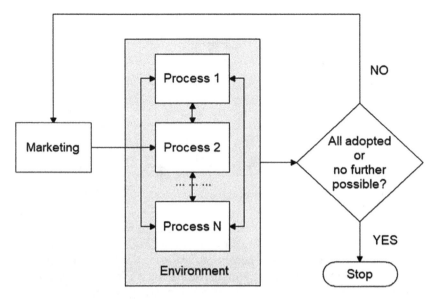

Figure 2.5 Representation of the system dynamic model as an agent-based model

How Do We Model a System?

When we consider modelling a system, where do we start? What should we take into account?

The model building process or the modelling process can be divided into four main phases: building, analysing, testing and using. While we could think of this as a linear progression (we build the model, analyse it, test it and then use it), more frequently these phases interact and loop. In fact, it might be necessary to update the original model (building phase) more than once, after analysing or testing, if issues are found during these phases, or if new information becomes available. Obviously, if the model changes, analysis and testing must be repeated (Figure 2.6).

Model building

When building a model, the obvious place to start is with the problem we want to solve and the system we need to represent. In other words, we need to clearly define the purpose of the model as it will influence the type and amount of information (data, typically) required. The information needed varies according to the objective. The most important features need to be determined, which are the variables that are fundamental to the problem we are trying to solve (i.e. they are key in the determination

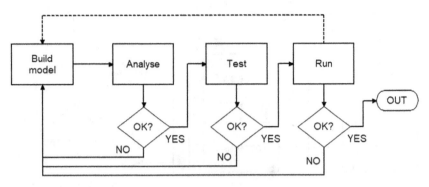

Figure 2.6 Representation of the model building process

of the model outcomes). To continue with the previous example, if we want to model a new marketing strategy aimed at increasing occupancy, we first need to reflect on how accommodations make decisions. Here, we may expect that the cost of implementation as well as the expected results play a key role in adopting the strategy. However, we may also anticipate that individual accommodations may look at the experience of their 'peers' and evaluate the success of early adopters of the strategy, as this increases confidence in the specific strategy and reduces the uncertainty of expectations. Hence, at a minimum, we would need to collect information or assume specific cost structures for the current occupancy, and how such changes may increase costs as well as how they would modify occupancy. Of importance is the network defined by knowledge sharing between processes existing between the different actors (see Chapter 3). In addition to this information, given the objective of the model, we may need to make some general assumptions about the environment. For example, we may want to assume only one new marketing strategy (the one we want to evaluate) rather than multiple different strategies, and we may also want to assume that differences in the location, size and quality of the accommodation are included in the cost and expected occupancy increase parameters. That is, simplifications must be made to keep the model tractable.

It is important to understand that assumptions are often made sequentially, that is one should first make assumptions on the fundamental variable that may affect the outcome of interest. These assumptions should always be made based on the relevant literature and/or empirical findings. Second, one needs to assume how parameters and variables interact, that is one needs to assume specific rules of interaction between the model component. Again, these assumptions should always be based on the relevant literature and/or observed behaviours/interaction. If none is available, one may need to 'experiment' with different rules to better assess the conditions under which the outcome of interest is influenced by variables, parameters and the interaction between them and the model components.

Modelling assumptions can be generally divided into two broad categories: quantitative assumptions and qualitative assumptions. Quantitative assumptions are the actual values (and equations) used to model interactions between the model components, and qualitative assumptions are associated with the *shape* of such equations, as well as with the general rules that govern the model itself. The final model, independently of whether it is a statistical, conceptual, agent-based or system dynamic model or it is based on machine learning methods,[3] is the result of the assumptions made. After all, a model is only as good as its assumptions.

To reiterate the concept, a model is something that represents the *reality* for a specific purpose, and thus every time we model, we tend to 'programme' or impose specific behaviours according to specific rules. This may raise several questions about models' semantics (what is the representational function that models perform?), ontology (what kind of objects are models?), epistemology (how do we learn with models?) and in the general philosophy of science (how do models relate to theory?; what are the consequences of a model-based approach for discussions on realism, reductionism, explanation and laws of nature?). We do not discuss these problems here (a succinct discussion can be found in Frigg & Hartmann [2018]), we only assume that a model is an abstract representation of a real-life situation.

The rules and programmes of a model allow the different components to evolve, and in some cases learn and adapt (Baggio, 2017; Bonabeau, 2002; Gilbert & Terna, 2000). Hence, the role of the modeler is that of defining the entities at play, and the ways in which the dynamics evolve (i.e. the interactions of the components over time) (Baggio, 2017; Galán *et al.*, 2009). More specifically, models should be built in three main steps:

(1) Conceptualisation of the system that needs to be modelled, including the purpose of the model (research question) and the identification of critical variables and their functional relationships.
(2) Formalisation of the conceptualised model.
(3) Coding (whether via algorithm or equations) and implementing the formalised model.

To summarise, when building a model, a modeler will first (a) determine the purpose of the model, then (b) assess which assumptions will need to be made and (c) assess the interaction between the model component and formalise protocols for such interactions (i.e. via equations, diagrams, algorithms).

Once the model is built, we need to analyse it. This step often includes assessing how the initial parameter configuration (the initial values based on our assumptions) affects the outcome of interest, and how varying such parameters affects the values of interest. This is often called *sensitivity analysis* and allows a modeler to judge how outcomes change depending on the initial parameter values chosen. The choice of *parameter space* (the complete set of parameters) to explore (i.e. the value range we should test) should, once again, depend on the system we are modelling, and thus on the relevant literature, observations, data collected and expert knowledge. In the absence of these, or when these sources cannot be considered reliable, a rule of thumb is to vary parameters by ±10% if we have a certain degree of confidence in the parameter values, or ±50% if the level of uncertainty for the different (usually the most important) parameter values is high.

Finally, once the model is built and analysed, an evaluation and validation phase follows. This can be a daunting task, especially when complex systems are at play. In an ideal situation, evaluation and validation are performed using actual data (if available), and the assessment consists of appraising how the model behaviour reproduces behaviours and outcomes empirically observed. Unfortunately, this type of evaluation can be hard to achieve due to cost and time constraints as well as the potential unavailability of data. The latter is especially true if we are to evaluate models that propose new strategies, policies or rules that have not been dealt with previously. Hence, we should evaluate a model based on how well our assumptions and interactions between the model components correspond to what actually happens (i.e. observed in reality). Although processes that determine the results of interactions in the real world are not always discernible, often it is possible to gauge whether the processes presented in the model are based on valid assumptions or not. Finally, if all else fails, we should ensure that the model is implemented correctly and, for example, no coding or mathematical (formal or logical) errors exist. The analysis and the outcomes of the model analysis phase should be solely the results of crafted assumptions, and not due to errors in the equations or algorithms implemented.

Besides evaluating a model *per se*, especially when the model objective is to predict the future, we need to judge the validity of the predictions that can be made. In fact, even when a model is implemented correctly, its assumptions have been validated empirically or grounded in the literature, and the rules, procedures and equations chosen to represent the model interactions are correct, a model can still fail to correctly predict outcomes.

In order to validate the ability of a model to correctly predict the future, the first action is, once again, to make sure that the parameters used and the assumptions made during the model building process are validated empirically or grounded in the literature. If this is the case, we may want to pay attention to the natural variability that exists in the system to be modelled. However, if we are modelling a system where we know that the natural variability is not a factor, we should reassess the model assumptions and the fundamental variables and parameters used. A common cause of misprediction is disregarding the fundamental factors, either because the modeler was unaware of them, did not correctly research them, or because such factors may have been unknown. In this latter case, the power of modelling is clear: it allows us to detect or suggest factors that may be key in influencing specific outcomes that were previously unidentified. In other words, a model can be used as a research initiator (another often unappreciated value of modelling).

Overall, it is worth highlighting the importance of the conceptual accuracy required to build models that are able to advance our theoretical understanding of a system or are able to predict (at least probabilistically) future outcomes. Every part of the model should be grounded in the literature or informed by experts (e.g. empirical researchers and practitioners), and the final test of any model is its importance in advancing the understanding and development of new formal theories as well as reducing the uncertainty that exists within a system related to specific decisions that need to be made.

Computational Models and Simulations

The simulation of a system is the operation of a model, typically in terms of time, for analysing the functioning of an existing or planned system. Simulation is thus the process of using a model to examine the performance of a system.

As already stated, it is important to recall that any model is only a limited representation of reality. That is, a model is an abstraction of essential features in relation to a specific purpose. Simulations (or computational models) can be employed to formalise our reality representations by retaining a certain level of complexity and, by avoiding oversimplification or overly strong assumptions, they imitate processes (Hartmann, 1996a, 1996b). Hence, via simulations it is possible explore different assumptions, hypotheses and parameters and provide insights into the realm they represent and describe, build scenarios or reach new theoretical developments (Garson, 2008; Hartmann, 1996a). Simulations allow us to assess and identify the dynamics of real processes, especially when such dynamics and processes cannot be rigorously identified via controlled empirical experiments (Hartmann, 1996a). A simulation model describes how the values of variables change during each iteration of the simulation.

Simulations are a powerful tool if used correctly, and as with any type of modelling, effort should be devoted to reflecting upon the assumptions made in order to represent the object of study. It is crucial to understand the role of assumptions in the model building process. Every equation, parameter, rule, inclusion or exclusion of variables is based upon certain hypotheses, and a model's validity is only as good as its assumptions (Silvert, 2001). The primary role of a researcher should be to identify and understand the implications of such assumptions. Every model, especially when seeking to represent complex systems, needs to be built through a process of continuous interactions between modelers and researchers or practitioners who deal with empirical issues. It is vital to understand what is happening in the field and how case studies, experiments and other techniques are employed (Peck, 2008; Silvert, 2001).

If built correctly, simulations can aid in better understanding the potential consequences of specific decisions, helping us to overcome the limitations of our cognitive abilities. In other words, simulations can help us overcome the computational, processing and classification limits of human cognition, especially when dealing with complex systems and large amounts of data (Durán, 2018).

Computational models, and more generally, models that represent complex interactions are often difficult to understand in detail and assessing their validity can be quite an intricate exercise. There is scepticism around computational models, as the results might be counterintuitive (though counterintuitive does not necessarily suggest incorrect). However, it is important to remember that not all models have as their key objective 'to predict'. In some cases, a model's main purpose is to allow for new theoretical developments and advances or to explain reality. If the model is considered plausible (within reason) and implemented correctly (with either the correct equations or correctly coded algorithms), it can still assist in advancing existing theories or deepening our theoretical understanding of the system under study even if its results might seem counterintuitive.

At the same time, it is important to be aware of the 'black box risk'. That is, the risk that the results might be the consequence of an unknown process inside a 'black box' (i.e. the computerised model) used to perform the simulation (Macy & Willer, 2002). Hence, when dealing with computational models it is key to publicise (i.e. report in detail) the models' code to the scientific community so that it will be possible to validate and, more importantly, replicate the results. Finally, a further important issue that needs to be tackled when devising, implementing and analysing a computational model is the fact that the value of such models for theoretical development could be dismissed as 'muddying the waters', as the number of variables and parameters and their relations may approach the complexity found in the real world (Peck, 2008).

Currently, no formal methodological procedure for building a computational model exists, although there are certainly similarities across all model building methods. As explained earlier, once the model is conceptualised and the parameters and variables are chosen as well as the way they will interact, the first step that needs to be considered is to ensure that there are no discrepancies between what one thinks the model is representing and what the coded model is actually doing (Galán et al., 2009).

A model must serve a purpose, and hence, must contain the right level of detail. As already noted, a model cannot retain all the real-world's details, thus it should be a simplified, although meaningful, representation of the system or phenomenon under investigation (Axelrod, 2006; Baggio, 2017). When constructing a model, it is necessary to abstract from the real world, hence when dealing with computational models, more so than when using other modelling techniques, it is necessary to refer to practitioners or draw on empirical research or carefully review the existing literature in order to gain insights into processes and fundamental behaviours that characterise the different system components and their interrelations. The absence of a clear research question will render the model less useful (or useless) in understanding the part of reality under investigation.

Albeit important for the modelling technique one wants to implement, from conceptual to statistical, from machine learning to networks, from system dynamic models to ABMs and more generally computational models, assumptions need to be thoroughly identified and the impact of each assumption on the results produced by the model should be carefully assessed. Given the complexity of such models, it is important to reiterate than one should be careful in assessing and looking for clear predictions and insights into the quantitative behaviour of computational models. The most important value that such models add is their ability to explain different configurations arising from the set of parameters used (i.e. scenarios), allowing a qualitative understanding of the system studied.

Simulations need to be approached and treated differently from traditional analytical models (Peck, 2008). The most challenging aspect of computational models resides in a careful understanding and planning of how entities in the model behave and interact with each other, which is especially true for system dynamic models as well as ABMs (as explained further in Chapter 3). The choice of the set of rules and feedback that relate different models is the central issue. It is vital to avoid assumptions that are not confirmed by 'general wisdom' (existing literature, experts, assessments, etc.). Therefore, as already emphasised, continuous interaction and feedback between researchers and 'experts' are necessary, so that it may be possible to shed light on the appropriate parameter space region to explore and the interactions that exist between the different model components.

Even when engaging continuously with experts and carefully planning a simulation following all good practices in the model building process, there is still room for errors and artefacts (Galán *et al.*, 2009). Here, by errors we mean a disparity between the coded model and the model intended to be coded (e.g. the modeler wants the model to call for Task A before Task B, but the model runs Task B before Task A). It is important to highlight the fact that there is no error if there is no disparity between the actual model and what was meant by the researcher, thus it is not possible to assert that an error exists if the modeler's objectives are not known. Intentions should always be clearly stated.

Artefacts, on the other hand, are disparities between the assumptions made by the researcher and thought to be the cause of specific results and what is actually causing them. This can happen because sometimes it is necessary to formulate hypotheses that are not critical for the system's representation but are nevertheless required in order to run the simulation code (e.g. in a spatial model, the size of the 'model space grid' might influence the results although its size is not one of the critical assumptions of the modelled system). In some cases, an artefact ceases to be an artefact when it is detected, and the cause of the results become known.

Both errors and artefacts can be avoided. In order to avoid errors, it is necessary to meticulously check the coding procedure and all its parts to ensure that the coded model is performing exactly as intended. Artefacts can be avoided by implementing a model with the same critical hypotheses but with different 'computer- or program-based' assumptions to control how results are affected (i.e. grid size, decimals used, floating-point arithmetic issues, etc.). This is a common procedure to assess the validity of the outcomes.

Evaluation of computational models and simulations

Often, the revealed behaviours of simulations are not initially understood (Srbljinovi & Škunca, 2003). Nonetheless, it is possible to evaluate simulations. The first criterion is an assessment of the model's reliability by allowing for different separate implementations and comparing the results. In other words, if time and resources allow, it is good practice to implement the model on different machines on different platforms or even code the model in different programming languages.

However, this alone is not sufficient to evaluate simulations and computational models that rely on peculiar algorithms. Three questions facilitate such an evaluation:

(1) Are the results similar to those observed in the real world, at least for some parameter configurations? Do these parameter configurations make sense? That is, do the parameter configurations that reproduce similar results to those observed in the real world make sense or are they impossible?
(2) Are the processes by which the different model components interact based on reliable assumptions and thus represent the processes observed in the real world? In other words, does the current state of the literature or empirical observation or expert judgement concur that the processes modelled are representative of the processes observed in the real world?
(3) Is the model coded correctly so that it is possible to state that the outcomes are the sole result of the model assumptions? That is, are the algorithms coded to represent the processes we are interested in correct? Are there errors and/or artefacts in the model code?

Answering the first two questions allows for assessing the validity of the representation (model), thus gauging how well the real system we want to describe is captured and explained by its representation. Answering the third question guarantees that the model's behaviour is what the modeler intended it to be (Galán et al., 2009).

As we will see in the next section and Chapter 3, the evaluation of a computational model can require an enormous amount of data that is not always available. Thus, one can resort to involving experts (communities, practitioners, other colleagues, literature available) to provide insights into the 'real' processes and dynamics and help evaluate the model's ability to represent reality. Moreover, it is worth highlighting the importance of the conceptual accuracy needed to build a computational model that is able to advance the theoretical understanding of a system. Computational models, and ABMs more so, explain rather than predict reality, allowing for a qualitative understanding of the fundamental processes underlying the system modelled.

Calibration, fitness and sensitivity analysis

Calibration of computational models and simulations can mean two different things. On the one hand, one can calibrate the model, that is, adjust the parameters of interest that depend on the model purpose (i.e. the research question), or assess via expert knowledge and calibrate the main processes observed. This calibration is often based on the values of specific parameters or a detailed description of processes found in the literature or empirically observed. In other words, this type of calibration process concerns finding and inputting the correct parameter values or coding the correct processes for the system under study. For example, with respect to reproducing processes, based on the work of Easterlin (2001, 2003), Baggio and Papyrakis (2014) analysed the effect of income distribution and income growth on happiness via a computational model. Schoon et al. (2014) based their model on observed patterns of species migration and management decision-making. Baggio and Hillis (2016, 2018) built their computational model to explain why farmers and, more generally, managers are locked in to less than optimal situations when it comes to managing complex issues.

On the other hand, with respect to parameter values, one can refer to the work of Balbi et al. (2013, 2015) who calibrated a model simulating the effects of climate change on tourism in the alpine region based on empirical data, and assessed the results of their model with local stakeholders; the model of Johnson and Sieber (2011) that allows the generation of different tourism visitation dynamics scenarios in a destination and simulates the effects of several actions; and the work of Shanafelt et al. (2017) whose model reproduces parameters for specific species and their interactions.

Another possible avenue to calibrate a model is the *adjustment* of specific scenarios or parameters to real-world data. This part of the process can only be performed after having observed output data. For example, one could devise a model to assess how different individual behavioural strategies lead to specific tourist flows in a certain geographic area. In this case, instead of calibrating the initial parameters (unless data on individual preferences are available), it is possible to tailor the model to simulate different parameters' values until some behavioural strategy leads to the observed outcomes. An example is calibrating models on observed results and on observed parameters or the work on high-quality data stemming from behavioural experiments, where the models are based on theoretical considerations, but are confronted with observed behaviour in a laboratory setting (Baggio & Janssen, 2013; Janssen & Baggio, 2017).

Computational models often involve multiple variables and parameters encoded in different data types (e.g. nominal, continuous and categorical). The recent tremendous increase in the availability of large volumes of data, when included in the modelling process, makes it possible to attain a more precise calibration and definition of models in tourism research. Empirically based and calibrated computational models have the advantage of improving the predictability of specific policies and decisions in complex systems (Janssen & Ostrom, 2006).

To calibrate such models to real-world data, however, it is necessary to resort to general optimisation algorithms that do not assume specific distributions. Frequently, in complex models, data can be nominal, ordinal and, in some instances, scalable. Hence, the most popular techniques to calibrate computational models representing general natural processes are often referred to as *genetic* algorithms or *simulated annealing* algorithms. Both are numerical metaphors of specific processes: metallurgical (simulated annealing) and biological (genetic algorithm) and aim to find (or approximate) optimal solutions by calculating a fitness function, an objective function used to state how close a solution is to a specified objective (i.e. observed reality).

Simulated annealing is inspired by the metallurgic process of annealing in which a metal is heated and cooled at regular intervals to modify its crystal structure to achieve some desired characteristics (ductility or robustness). The numerical algorithm mimics this process to find a close-to-optimal (not necessarily the best) solution to the problem of optimising a large (but finite) set of potential solutions. It starts with a random set of parameter values and then considers a neighbouring state of that parameter configuration. It confronts the fitness of both parameter combinations and probabilistically chooses a configuration based on the difference between the two fitness measures (Kirkpatrick *et al.*, 1983; van Laarhoven & Aarts, 1987). This process is run until a specific fitness value is attained or a pre-set number of computation cycles is reached (i.e. a predetermined number of configurations is reached).

Genetic algorithms are based on natural selection processes. Parameter values are initiated and fitness is calculated, then new parameter configurations are created and new solutions are generated. The fitness of these new solutions is again calculated and the fitter ones are more likely to be *reproduced* for the next generation (Mitchell, 1996).

In short, the main difference between different search algorithms principally resides in the probability of accepting solutions (or parameter configurations) that may not be any better than the previous solutions. The probability of accepting 'worse' solutions can obviate the issue of being *trapped* in local optima: take a step back to take two forward, to avoid being 'stuck' in a suboptimal region of the parameter space. All these algorithms have in common the use of a fitness function necessary to discriminate better versus worse solutions. A fitness function can be maximised or minimised. For example, one could minimise the distance between actual and observed data. A fitness function should be monotonic, that is, always increasing or decreasing. If the fitness function is non-monotonic, the search algorithm may not lead to best (or approximately best) global optima.

Due to their probabilistic nature, none of these algorithms guarantees the detection of a global optimal solution. It is therefore suggested that several runs are made, each starting from a different initial condition. It is also possible to combine the two families, for example looking for a solution with a genetic algorithm and refining it to use as a starting point for a simulated annealing process (a recent example can be found in Junghans & Darde [2015]).

Once models are calibrated, it is necessary to perform a sensitivity analysis, which is key to assessing how changes in individual parameters affect the outcomes of a model (ten Broeke *et al.*, 2016). Many methods can be used to perform a sensitivity analysis, though preference should be given to the exploration of the plausible parameter space, either by varying one parameter at a time or by co-varying parameters. One should be cautious of performing any type of statistical analysis on models until the generating function that produces the outcomes is known (it is modelled) (Smaldino, 2016). The most commonly used methods are changing one parameter at a time or performing a global sensitivity analysis (ten Broeke *et al.*, 2016).

Global sensitivity analysis centres on the relationship between model outcomes and a wider range of parameter values and configurations, normally drawn from specific distributions. One factor at a time, on the other hand, means changing individual parameters by a specific amount, while keeping all other parameters constant. One factor at a time allows an understanding of how single parameters influence model outcomes. In the tourism context, for example, sensitivity analysis can be used to assess the most important factors influencing tourists' movements, which can be crucial for the construction of scenarios able to support specific decisions or policies.

Concluding Remarks

This chapter has provided an overview of modelling activities, discussing what models are, why they are important and how to build, analyse and validate a model. All models share common features: (a) they are a simplified representation of something; (b) they follow a messy but important building process in which they are conceptualised, built, implemented, analysed and, if required, amended; and (c) their evaluation, especially when they are to represent complex systems, is not straightforward. To this purpose, we can summarise the important verification, validation and testing activities using the guiding principles discussed by Balci (1998):

- verification and validation activities must be conducted throughout the entire modelling and simulation life cycle;
- the outcome of verification, validation and testing should not be considered a binary variable where the model or simulation is absolutely correct or absolutely incorrect;
- a simulation model is built with respect to stated objectives, and its credibility is judged with respect to those objectives;
- verification and validation require independence to prevent developer's bias;
- verification, validation and testing are difficult and require creativity and insight;
- credibility can be claimed only for the prescribed conditions for which the model or simulation is verified, validated and accredited;
- complete simulation model testing is not possible;
- verification, validation and testing must be planned and documented;
- type I, II and III errors must be prevented;
- errors should be detected as early as possible in the modelling and simulation life cycle;
- multiple response problems must be recognised and resolved;
- successfully testing each submodel (module) does not imply overall model credibility;
- double validation problems must be recognised and resolved;
- simulation model validity does not guarantee the credibility and acceptability of simulation results;

- a well-formulated problem is essential to the acceptability and accreditation of the modelling and simulation results.

Further, we have also treated issues specifically related to more complex models and the importance of computational models and simulation as a cognitive aid for managers and developers, as well as issues arising from the willingness to represent complex phenomena. Finally, we have reiterated the importance of following very detailed protocols with respect to documenting a model, making sure that all assumptions are thoroughly vetted (observed reality and primary data collection, literature, expert knowledge, etc.).

The whole approach is schematically summarised in Figure 2.7.

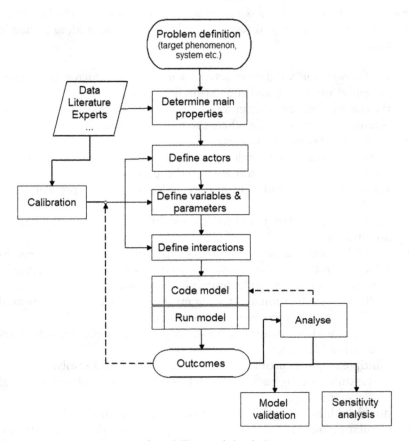

Figure 2.7 A schematic view of modelling and simulation

In Chapters 3 and 4, a detailed overview is provided of different modelling techniques (from conceptual to statistical, from machine learning to networks to ABMs), and Chapter 5 details how to choose an appropriate modelling technique.

Notes

(1) A description of these techniques is provided in Chapter 3.
(2) See Chapter 3.
(3) These types are discussed in Chapter 3.

References

Axelrod, R. (2006) Advancing the art of simulation in the social sciences. In J.-P. Rennard (ed.) *Handbook of Research on Nature Inspired Computing for Economy and Management* (pp. 90–100). Hershey, PA: Idea Group Reference.

Baggio, J.A. (2017) Complex adaptive systems, simulations and agent-based modeling. In R. Baggio and J. Klobas (eds) *Quantitative Methods in Tourism: A Handbook* (2nd edn) (pp. 223–244). Bristol: Channel View Publications.

Baggio, J.A. and Janssen, M.A. (2013) Comparing agent-based models on experimental data of irrigation games. In R. Pasupathy, S.-H. Kim, A. Tolk, R. Hill and M.E. Kuhl (eds) *Proceedings of the 2013 Winter Simulations Conference* (pp. 1742–1753). Piscataway, NJ: IEEE Press.

Baggio, J.A. and Papyrakis, E. (2014) Agent-based simulations of subjective well-being. *Social Indicators Research* 115 (2), 623–635.

Baggio, J.A. and Hillis, V. (2016) Success-biased imitation increases the probability of effectively dealing with ecological disturbances. In T.M.K. Roeder, P.I. Frazier, R. Szechtman, E. Zhou, T. Huschka and S.E. Chick (eds) *Proceedings of the 2016 Winter Simulation Conference* (pp. 1702–1712). Piscataway, NJ: IEEE Press.

Baggio, J.A. and Hillis, V. (2018) Managing ecological disturbances: Learning and the structure of social-ecological networks. *Environmental Modelling & Software* 109 (August), 32–40.

Balbi, S., Giupponi, C., Perez, P. and Alberti, M. (2013) A spatial agent-based model for assessing strategies of adaptation to climate and tourism demand changes in an alpine tourism destination. *Environmental Modelling and Software* 45, 29–51.

Balbi, S., del Prado, A., Gallejones, P., Geevan, C.P., Pardo, G., Pérez-Miñana, E., Manrique, R., Hernandez-Santiago, C. and Villa, F. (2015) Modeling trade-offs among ecosystem services in agricultural production systems. *Environmental Modelling and Software* 72, 314–326.

Balci, O. (1998) Verification, validation, and testing. In J. Banks (ed.) *Handbook of Simulation* (pp. 335–393). New York: Wiley.

Bonabeau, E. (2002) Agent-based modeling: Methods and techniques for simulating human systems. *PNAS* 99 (suppl. 3), 7280–7287.

Borshchev, A. and Filippov, A. (2004) From system dynamics and discrete event to practical agent-based modeling: Reasons, techniques, tools. In M. Kennedy, G.W. Winch, R.S. Langer, J.I. Rowe and J.M. Yanni (eds) *Proceedings of the 22nd International Conference of the System Dynamics Society* (pp. 25–29). Albany, NY: System Dynamics Society.

De Santillana, G. and Von Dechend, H. (1977) *Hamlets Mill: An Essay on Myth and the Frame of Time*. Boston, MA: David R. Godine.

Durán, J.M. (2018) *Computer Simulations in Science and Engineering*. Cham: Springer International Publishing.

Easterlin, R.A. (2001) Income and happiness: Towards a unified theory. *Economic Journal* 111, 465–484.

Easterlin, R.A. (2003) Explaining happiness. *Proceedings of the National Academy of Sciences* 100, 11176–11183.

Foresman, P.S. (2008) Archives of Pearson Scott Foresman: Donated to the Wikimedia Foundation. See https://commons.wikimedia.org/w/index.php?curid=3947167 (accessed March 2019).

Frigg, R. and Hartmann, S. (2018) Models in science. In E.N. Zalta (ed.) *The Stanford Encyclopedia of Philosophy* (summer 2018 edition). See https://plato.stanford.edu/archives/sum2018/entries/models-science/ (accessed December 2018).

Galán, J.M., Izquierdo, L.R., Izquierdo, S.S., Santos, J.I., del Olmo, R., López-Paredes, A. and Edmonds, B. (2009) Errors and artefacts in agent-based modelling simulation of social processes. *Journal of Artificial Societies and Social Simulation* 12 (1), 1. See http://jasss.soc.surrey.ac.uk/12/1/1.html (accessed December 2018).

Garson, G.D. (2008) Computerized simulation in the social sciences: A survey and evaluation. *Simulation & Gaming* 40 (2), 267–279.

Gilbert, N. and Terna, P. (2000) How to build and use agent-based models in social science. *Mind & Society* 1 (1), 57–72.

Hartmann, S. (1996a) The world as a process: Simulations in the natural and social sciences. In R. Hegselmann, U. Mueller and K.G. Troitzsch (eds) *Modelling and Simulation in the Social Sciences from the Philosophy of Science Point of View* (pp. 77–100). Berlin: Springer Science & Business Media.

Janssen, M.A. and Ostrom, E. (2006) Empirically based, agent-based models. *Ecology and Society* 11 (2), art. 37.

Janssen, M.A. and Baggio, J.A. (2017) Using agent-based models to compare behavioral theories on experimental data: Application for irrigation games. *Journal of Environmental Psychology* 52, 194–203.

Johnson, P.A. and Sieber, R.E. (2011) An agent-based approach to providing tourism planning support. *Environment and Planning B: Planning and Design* 38 (3), 486–504.

Junghans, L. and Darde, N. (2015) Hybrid single objective genetic algorithm coupled with the simulated annealing optimization method for building optimization. *Energy and Buildings* 86, 651–662.

Kirkpatrick, S., Gelatt, C.D. and Vecchi, M.P. (1983) Optimization by simulated annealing. *Science* 220 (4598), 671–680.

Macy, M.W. and Willer, R. (2002) From factors to factors: Computational sociology and agent-based modeling. *Annual Review of Sociology* 28 (1), 143–166.

Mitchell, M. (1996) *An Introduction to Genetic Algorithms*. Cambridge, MA: MIT Press.

Peck, S.L. (2008) The hermeneutics of ecological simulation. *Biology and Philosophy* 23 (3), 383–402.

Poincaré, H. (1884) Sur certaines solutions particulières du problème des trois corps. *Bulletin Astronomique* 1, 63–74.

Schoon, M., Baggio, J.A., Salau, K.R. and Janssen, M. (2014) Insights for managers from modeling species interactions across multiple scales in an idealized landscape. *Environmental Modelling and Software* 54, 53–59.

Shanafelt, D.W., Salau, K.R. and Baggio, J.A. (2017) Do-it-yourself networks: A novel method of generating weighted networks. *Royal Society Open Science* 4 (11), 171–227.

Silvert, W. (2001) Modelling as a discipline. *International Journal of General Systems* 30 (3), 261–282.

Skeat, W. (ed.) (2014) *Chaucers Works, Volume 3*. Project Gutenberg. See http://www.gutenberg.org/ebooks/45027 (accessed January 2018).

Smaldino, P.E. (2016) Models are stupid, and we need more of them. In R.R. Vallacher, S.J. Read and A. Nowak (eds) *Computational Models in Social Psychology* (pp. 311–331). New York: Routledge.

Srbljinovi, A. and Škunca, O. (2003) An introduction to agent-based modelling and simulation of social processes. *Interdisciplinary Description of Complex Systems* 1 (1–2), 1–8.

ten Broeke, G., van Voorn, G. and Ligtenberg, A. (2016) Which sensitivity analysis method should I use for my agent-based model? *Journal of Artificial Societies and Social Simulation* 19 (1), art. 5.

van Laarhoven, P.J.M. and Aarts, E.H.L. (1987) *Simulated Annealing: Theory and Applications*. Dordrecht: Springer Netherlands.

3 Methodological Approaches

Introduction

This chapter describes the most relevant and most used modelling methods. Many models serve different purposes and consider different aspects of a system or different types of systems. A broad classification can be useful in guiding the researcher in selecting the right type of model for achieving the scope set. Although distinctions are not always clear, we can roughly divide models into two families: descriptive and analytical.

A descriptive model aims to provide a description of the logical relationships between the different parts of a system, and the functions that these components perform. Typical examples are those that depict the functional or physical architecture, or the role that components play in the production of outcomes or in the modifications of a system's behaviours.

An analytical model uses mathematical relations for a quantitative analysis of the main parameters and the effects they have on what the system may produce or in how the system configuration is achieved. They can study a static picture or the dynamic, time-evolving state of a system. For example, a dynamic model may characterise the performance or the efficiency of a system when its main parameters are modified, while a static model could represent an estimate of the general properties at a certain time or corresponding to a certain state.

Independently of the general type, a model can focus on different aspects such as (see e.g. Cloutier, 2017):

- properties of the system, such as performance, reliability, mass properties and power;
- design and technology implementations, such as electrical, mechanical and software design models;

- subsystems and products, such as communications, management and power distribution models;
- system applications, such as information systems, automotive systems, aerospace systems and medical device models.

The methods presented here concern conceptual models (CMs), statistical and analytical models, machine learning (ML), system dynamics (SD), network models and agent-based models (ABMs). We present them using a single case, so that it is clearer how the different approaches contribute to the general knowledge of the system under study and what aspects the different techniques address.

The object of study

The system used as an example is an Italian tourism destination: the city of Cremona and its surroundings. Cremona is a city in northern Italy, located in Lombardy, on the left bank of the Po River in the middle of the Po Valley (*Pianura Padana*) (see Figure 3.1). It counts about 70,000 inhabitants. The city has been a famous music centre since the 16th century, being home to renowned families of luthiers such as the Amati family, Giuseppe Guarneri and Antonio Stradivari. This tradition continues and Cremona is still known for its artisan workshops producing high-quality string instruments.

Cremona has an attractive historic centre with many medieval and Renaissance monuments and rich museums, such as the world-famous violin museum. Most sights are clustered around the main square, Piazza

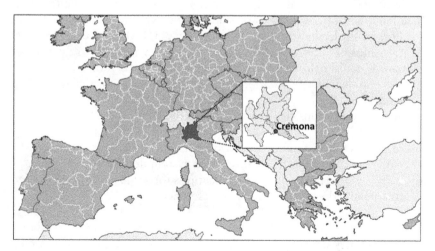

Figure 3.1 The city of Cremona, Italy

del Comune, considered one of the most beautiful medieval squares in Italy. The city also offers a rich gastronomic heritage with authentic products and is an important cultural centre, hosting numerous art exhibitions.

These characteristics make Cremona an attractive tourism destination with about 200 accommodation facilities (about 4,000 bed places), receiving almost 250,000 tourists (mostly domestic) per year who spend, on average, 2 days at the destination.

Most of the data used in what follows come from an unpublished MSc thesis (Milo, 2015) that analysed and discussed many aspects of the destination system, from the yearly data collected by the Italian Statistical Bureau (ISTAT, available at: http://dati.istat.it/), and the Bank of Italy's border survey (Indagine sul turismo internazionale dell'Italia) on inbound tourism to Italy (available as aggregate values and microdata at: https://www.bancaditalia.it/pubblicazioni/indagine-turismo-internazionale/index.html). These are based on interviews and report counting of resident and non-resident travellers at the Italian borders, the number and the name of the locations visited and the satisfaction expressed for a number of activities performed at the destination.

Conceptual Models

A CM is a representation of a system essentially composed of idealised concepts or ideas. The objective is to help people know and understand or simulate a subject represented by the model. CMs are used as an abstract description of real-world objects or systems in a wide range of disciplines, and usually form the basis for further works such as more detailed quantitative analyses or simulations (Robinson, 2008a, 2008b).

For example, in software development, a CM may exemplify the relationships of entities within a database, and when used in the tourism domain it could help in understanding the power relationships between the stakeholders of a destination or those factors that affect its image (e.g. Gallarza *et al.*, 2002).

A good CM fulfils the following objectives:

- improves understanding of the system considered;
- eases the communication of relevant system details;
- provides a reference when specifications for some process are needed;
- documents the system for future reference.

For example, a good model for a tourism destination structure should capture the key entities (companies, associations, groups) and the most relevant relationships between them.

Successful modelling takes into account the following steps:

- understands the problem situation;
- defines the relevant terms and entities used in the model;
- determines the modelling and general project objectives;
- identifies the model inputs and outputs;
- finalises content (scope, details) and identifies assumptions and simplifications.

Moreover, a CM is a dynamic entity that can be changed when information about a system is updated or deeper reflections and considerations lead to different elements or changes in the relative importance. Normally, a CM is expressed with visual and written diagrams that can quickly convey the abstract concepts presented.

The source for a CM is usually a qualitative investigation, complemented by a thorough scan of the previous literature on the subject. At a later stage, a CM can be used to inform some more complex modelling technique or some empirical quantitative work aimed at verifying the reliability and the correctness of the conceptual construction. For example, through a series of unstructured interviews, Kim *et al.* (2009) sketched a CM to determine the factors that influence the consumption of local food and beverages in a destination. Kim *et al.* (2009: 423) identified three main categories: '"motivational factors" (i.e. exciting experience, escape from routine, health concern, learning knowledge, authentic experience, togetherness, prestige, sensory appeal, and physical environment); "demographic factors" (i.e. gender, age, and education); and "physiological factors" (i.e. food neophilia and food neophobia)'.

In a later work, Kim *et al.* (2013) used this model as a basis for an empirical survey conducted at a number of destinations. The survey allowed the authors to confirm some of the relationships previously identified in the CM and to discard other possible connections. They found that some demographic factors, in particular, gender and age, remained as important variables, while no significant relationship was found between motivational factors and food-related personality traits (the physiological factors). Figure 3.2 shows the differences.

It is important to remark here that the CM had the critical function of guiding the choice of elements to study for the empirical work on consumption determinants. The first step is the formulation of a CM before starting a systematic investigation on some issue. Let us now consider the city of Cremona.

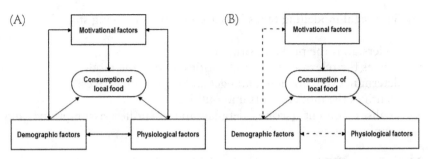

Figure 3.2 The conceptual mode of local food consumption (A) and the outcomes of its empirical verification (B). Dotted lines indicate partial relationship (Pictures adapted from Kim *et al.*, 2009, 2013)

A qualitative exploration (12 unstructured interviews among relevant individuals in the city) led to identifying the activities that mostly affect tourism in the city and the most relevant entities that manage or govern actions and strategies (Milo, 2015). Here 'activities' indicates all operators (companies, groups, individuals, etc.) that work in that particular segment in the city.

These entities are composed of public and private companies, organisations or associations. Local tourism systems are governance bodies established in areas of particular tourist interest (Italian law allows but does not 'force' regions to establish these bodies). Essentially, they are boards composed of all the relevant touristic, commercial, industrial

Table 3.1 A conceptual model for tourism activities and governance in Cremona

Activities	Entities						
	Local tourism system	Chamber of commerce consortium	Cremona consortium	Music district	Oglio Chiese district	Tourism board	Gastronomy consortium
Arts and culture	X	X	X	X	X		
String instruments	X	X	X	X		X	
Music	X	X	X	X		X	
Gastronomy	X	X	X		X		X
Events	X	X	X				X
Cremona fair	X	X				X	
Po River navigation	X	X	X		X		
Cycleways	X	X	X		X		

Source: Adapted from Milo (2015).

and cultural actors (or their representatives) of a territory that have the objective to put in place concerted strategies and actions for improving tourism activities in consultation and partnership with all interested parties. The other entities identified are typically associations or consortia (sometimes called 'districts') that handle specific segments (geographic, business, cultural, etc.). Table 3.1 summarises the most important relationships between activities and organisations.

This CM clearly shows the basic connections between the different actors in the destination and may allow a better examination of their common or competing interests, their activities and the relevance of their actions for the tourism phenomenon in the destination.

Statistical Models

Statistical models are used to extract relevant information from raw data. The domain of statistical techniques is vast, and their wide use has made available a large number of publications on a huge number of topics, from more general theoretical treatments to practical handbooks to quite specific issues. In the tourism and hospitality domain, the subject is typically treated in research methods books (e.g. Veal, 2006), but some dedicated texts exist (Baggio & Klobas, 2017; Smith, 2013). In the same way, many software applications are available, from those of a general use to those that deal with very specific issues (see e.g. the list available at en.wikipedia.org/wiki/List_of_statistical_packages).

Given the wide popularity of statistical methods, here we will limit ourselves to examining the most relevant issues concerning the task of statistically modelling a system or a process.

Typically, the starting point is a CM (whether formally described or informally sketched) of the system or the issue that we want to explore. In general, some assumption is made about certain properties, operationalised as a numeric variable, and on how this allows us to calculate the probability of an event, or to what extent this variable affects or is associated with some other properties. More precisely, a statistical model is a family of probability distributions, a combination of inferences based on collected data, used to evaluate error estimates of observations, relationships between variables, their grouping or to predict information on the future behaviour of a system based on its past history and allow extrapolation or interpolation of data based on some best fit. A model starts with the definition of a hypothesis that is then evaluated quantitatively with a mathematical relationship. A statistical model can be seen as a formal representation of a theory or, in abstract form, of the process that generated the data examined. Moreover, in many cases, different methods are used to achieve the same (or similar) outcomes and comparisons are made to assess the similarity, and hence the validity, of the various conclusions (Konishi & Kitagawa, 2008).

Besides understanding the relationships between different aspects (or system elements), statistical models can provide some predictive capabilities, although often limited in some respects, that are useful for planning or design activities, or as a basis for building future scenarios or even for detecting some type of unknown event, regardless of when it occurred.

Once all the data have been collected, the first step is a univariate descriptive analysis of all the variables involved. This allows a better understanding of the type of modelling procedures that can be applied in the case studied (e.g. parametric or non-parametric procedures) and to better assess the overall quality of the data collected, such as checking for outliers or missing values, or identifying other anomalies, noisy values or systematic measurement errors. Moreover, a test on possible collinearities in the variables can be performed to assess possible associations and correlations that can be detrimental in certain procedures (e.g. in a multiple regression). In this phase, it is also possible to apply some transformation to the collected values (normalisation, standardisation, etc.), rescaling values that could be of very different 'size' (depending on the measurement units), thus negatively affecting many of the statistical procedures, or to prepare the data for specific techniques (discretising, dichotomising, etc.) (García *et al.*, 2015).

Then, based on the number of variables examined, a statistical model can use a wide range of procedures aimed at verifying or disproving the hypothesis set. The most common procedures involve:

- factor and principal component analyses to identify patterns in the correlations between variables, to infer the existence of underlying latent variables (factors, components and dimensions) or to recognise variables that are composites of the observed quantities;
- various types of regressions (single, multiple, linear or non-linear) in which the objective is to deduce a mathematical relationship that quantitatively describes the strength of the relationship between a certain quantity (dependent variable) and a series of other changing variables (independent variables);
- conjoint analysis, which is a survey-based technique that is used to determine how people value the different attributes (features, functions, benefits) of an item;

- structural equation modelling that, using a combination of factor analysis and multiple regression, examines structural relationships between measured variables and latent constructs;
- time-series analysis methods that use temporal sequences of values to extract meaningful characteristics from the data with the main objective of forecasting future values based on historical observations.

As an example, we use the CM for tourism activities in Cremona (see Table 3.1) and set a model to assess the relative importance of these activities in determining the satisfaction of international visitors.

Data on this issue can be found in the Bank of Italy's border survey previously mentioned. The rich set of variables also contains an assessment of the opinions of visitors on a number of items and their overall satisfaction with their visit.

The items we consider are

- hospitality: people's courtesy, welcoming attitude, likeability;
- art: art cities and artworks;
- environment: landscape, natural environment;
- hotels: hotels and other accommodations;
- food: meals, cuisine, gastronomic products;
- prices: prices, cost of living;
- shopping: quality and variety of goods available in shops;
- info: information and tourist services;
- security: security of tourists.

We perform a multiple regression looking for the highest contributors to the overall satisfaction (dependent variable). Although not all these items are identical to those identified in the CM, many of them are compatible with the 'activities' reported in Table 3.1. For example, 'arts and culture', 'string instruments' and 'music' can be grouped in the 'art' variable. Here, however, we see that other components are at play, namely the accommodation sector that our CM had somehow neglected.

The analysis is conducted with SPSS (a specialised software for statistical analysis).

We recover 245 records from the full data set (people who have declared to have visited Cremona). The average values of the satisfaction expressed (on a scale 1–10, 1 = min) are shown in Table 3.2.

The model produced allows us to identify the variables that significantly contribute to the overall satisfaction: hospitality, art, hotels, food and shopping. The other factors (environment, security, info, prices)

Table 3.2 Opinions on Cremona

Item	Value
Overall	8.4
Hospitality	8.3
Hotels	8.2
Art	8.4
Food	8.7
Shopping	8.5
Environment	8.4
Security	8.3
Info	8.0
Prices	6.9

have little influence. The adjusted R^2 (which measures the goodness of the fit) is 0.710, a fairly good value. The other results are shown in Table 3.3.

First of all, we note that these are all statistically significant and that collinearity is not an issue. The variance inflation factor (VIF), one of the possible indicators usable in these considerations, has small values (values over 10 are considered problematic) (see e.g. Baggio & Klobas, 2017: 101).

The coefficients of the different variables give us an indication of the strength of each variable's effect on the overall satisfaction. We do this by looking at the standardised coefficients (in our case all variables are on the same scale so the unstandardised coefficients could also be used). SPSS provides the partial correlations between the variables chosen. Among these, the 'part' is the semi-partial correlation that shows (similarly to the coefficients) the contribution of each variable to the overall model: how much R^2 increases when the variable is added to a model that already contains all of the others.

Cremona is a small city, so it might be expected that the 'environment' variable has little effect on the overall model. The same can be said of 'security' and 'info' (information on the destination characteristics) variables, which have a good level of appreciation (see Table 3.2). It is also interesting to note that prices do not impact significantly, and that shopping (a model variable) has a negative effect on the overall satisfaction, probably due to the low evaluation of the prices (obviously considered too high).

With these results, we have, at least partially, validated our CM and are able to assign priorities to the possible actions directed towards the

Table 3.3 Regression model results

	Unstandardised coefficients		Standardised coefficients			Correlations			Collinearity statistics
	B	Std. error	Beta	Sig.	Zero-order	Partial	Part	VIF
(Constant)	1.405	0.609		0.023				
Hospitality	0.367	0.043	0.579	0.000	0.632	0.686	0.492	1.383
Hotels	0.235	0.051	0.315	0.000	0.473	0.452	0.265	1.418
Art	0.249	0.071	0.259	0.001	0.550	0.356	0.199	1.694
Food	0.183	0.058	0.239	0.002	0.606	0.327	0.181	1.735
Shopping	–0.208	0.077	–0.201	0.008	0.375	–0.284	–0.155	1.682

Note: Only columns useful to the discussion are reported.

main destination's components. Moreover, the finding that elements such as the accommodation sector significantly contribute to the overall opinion will prompt us to update the CM with this addition.

Machine Learning

In the last decades, the incredible development in the number and sophistication of information and communication technologies has brought, among many other effects, a huge proliferation of data, often in connection with user-generated content stemming from a vast array of online applications (Leung *et al.*, 2013), which have been made accessible through a number of mechanisms and tools. The features of the data available have also posed a challenge to the traditional analysis methods. These 'big data', as they are usually called, are characterised by a high degree of variety (i.e. wide range of forms and shapes, such as texts, sounds, pictures and videos), with great variability, in terms of meaning that can vary across contexts and times, velocity (the speed at which they are created) and, obviously, size, that can easily reach millions of gigabytes (Gandomi & Haider, 2015; Zikopoulos *et al.*, 2012). Moreover, often, whole populations instead of samples are gathered for a study.

All these properties make the analysis quite challenging and necessitate a reconsideration of many of the statistical tools and inferential methods usually employed (Fan *et al.*, 2014). In fact, the relational nature of variables, frequently common across different sources, and the flexibility needed when analysing collections that might involve extending variables and cases (Mayer-Schönberger & Cukier, 2013) or when higher levels of granularity are desirable, require tools that are different from those that statisticians have been developing for many years.

The great volume of variables that can be collected generates what is known as the 'curse of dimensionality', the exponential increase in the complexity, and the cost, of the analysis procedures, which makes the solution difficult, if not impossible, to find using almost any of the traditional numerical methods (Keogh & Mueen, 2011), or leads to erroneous or nonsensical outcomes (spurious correlations, inflated significance levels or overstated parameters' estimates; see e.g. Granville [2013]). Finally, no standard statistical method is able to properly handle non-numerical data (often in incompatible formats or structures) such as images, texts, sounds or videos, which can be too big to hold in a computer's memory, or when computing takes too long to be practicable (Wang *et al.*, 2016). This situation has attracted the attention of computer and data scientists, whose endeavours have been directed to novel techniques for data analysis (Franks, 2012).

Machine learning is a term that is used to define the methods and algorithms used for analysing data with the aim of extracting patterns, correlations and knowledge from apparently unstructured sources. It can be defined as the 'programming of a digital computer to behave in a way which, if done by human beings or animals, would be described as involving the process of learning' (Samuel, 1959: 211).

The idea at the basis of ML is that an algorithm can discover general rules or features in large data sets, 'learning' from the data without using explicit procedures, and is able to perform a task by using generalised approaches without being explicitly programmed for single jobs (Mitchell, 1997; Witten *et al.*, 2016). ML makes no prior assumptions about the possible relationships between variables. A specific algorithm, sometimes guided by examples, processes the data and uncovers configurations that can be used to make predictions on some data of interest.

The availability of powerful hardware and software libraries is making ML a popular method in addressing a number of problems and questions about the huge quantities of data available. Algorithms are seen as black boxes, and are generally applied to high-dimensional data sets. Although most of the basic mathematical techniques are quite 'old', the field is rapidly evolving and the possible ways of addressing an inquiry are rapidly multiplying. For example, the two most common reference works contain several thousand pages and close to 1000 entries each (Kao, 2016; Sammut & Webb, 2017).

There are several differences between a statistical approach and ML methods. Statistical procedures focus on the formalisation of the relationships between variables, expressed in terms of mathematical equations, to allow predicting outcomes or assessing structural differences or similarities or quantifying uncertainties. Usually, these procedures include the production of effect sizes or confidence intervals (Baggio & Klobas, 2017). ML, as said, aims at building systems that can learn from the data without explicitly specific programming. They are often quite intensive from a computational point of view and mostly deal with classification problems or predictive modelling. While statistical procedures require numeric quantities, ML algorithms can be fed with a vast and varied set of items, usually non-structured, or poorly structured or formalised (Bishop, 2006).

The broad series of ML algorithms can be roughly divided into two main families: supervised and unsupervised. The first are those that require 'training', a set of pre-labelled data.

Inputs and outputs are mapped onto a function using example input–output pairs. The software infers the basic features from the input data and derives a function able to reproduce the given outputs. The model built can then be used for mapping new items. The precision and accuracy of the results are assessed by splitting the training data set into two parts. The first part is used for the learning phase of the model that is then applied to the second part. A comparison between the original labels and those inferred by the algorithm gives a measure of the capacity to provide meaningful outcomes. Typical supervised algorithms are those that deal with classification or regression problems.

Unsupervised algorithms derive all needed features directly from the data and model their distributions to produce a model that can be applied to new instances. Clustering or dimensionality reduction issues are typically treated in this way.

The distinction, however, is not always clear and mixed forms exist. In between the two classes described, for example, we can find an

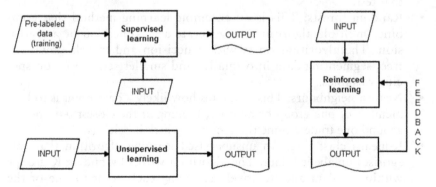

Figure 3.3 Schematic view of the main classes of ML techniques

ensemble of reinforcement learning algorithms. These may start with a supervised or an unsupervised learning procedure and produce some output that is then verified and validated (manually or algorithmically). The outcomes of the validation are then re-input into the system that eventually updates its previous conclusions, thereby improving the accuracy and the precision of subsequent results. Figure 3.3 provides a schematic view of ML types of algorithms.

A further class of algorithms are known as 'deep learning'. These will be discussed in Chapter 4.

The most known and used ML algorithms are:

- Naïve Bayes classifier: A supervised probabilistic algorithm that classifies every value as independent of any other value, based on a given set of features.
- Support vector machine: A supervised learning model for classification and regression analysis. It essentially filters data into categories, based on a set of training examples. The algorithm builds a model that assigns new values to one category or the other.
- Linear regression: The most basic type of regression. It produces a mathematical relationship between one variable and another or a set of other variables.
- Logistic regression: Despite its name, it is a classification algorithm that estimates the probability of an event occurring or not occurring based on the data provided. It is used to analyse a binary-dependent variable.
- Decision trees: A tree structure that uses a branching method to explain every possible outcome of a decision. Each node within the tree is a test on a specific variable, and each branch is the outcome of that test.
- Random forests: This is an ensemble learning method that combines multiple algorithms to generate a classification or a regression. The algorithm starts with a decision and travels down the tree, segmenting data into smaller and smaller sets, based on specific variables.
- Nearest neighbours: This estimates how likely a data point is to be a member of one group or another, looking at the closest data points around one to determine the group to which it belongs.
- K-means clustering: An unsupervised learning algorithm that categorises unlabelled data. The algorithm works by finding K groups within the data and iteratively assigning each point to one of the groups, thereby minimising the intra-cluster variance.

- Association rules: These allow analysing data for patterns or co-occurrences, determining the most recurrent 'if–then' associations after having identified the most frequent itemsets (groups of items that share common variable values). It is the basis for recommendation systems and is used in the analysis of shopping baskets.

In essence, an ML model is built using a scheme such as that shown in Figure 3.4. Starting from a research question with an unknown function to be found, we input data into an algorithm (adding several training examples if needed for a supervised algorithm) and derive outcomes that need to be carefully examined, interpreted and explained.

As an example, we use the Bank of Italy's border survey data in which, as previously shown, tourists are asked a number of questions, including the locations they visited during their trip. By selecting those that mention Cremona, we set out to identify the most frequent locations visited in the same trip in order to arrive at a set of association rules that enables us to better understand the paths followed by international visitors.

Several software applications provide this possibility. Here, we use Rapidminer (Mierswa & Klinkenberg, 2018). It is a software platform that provides an integrated environment for data preparation, ML, deep learning, text mining and predictive analytics. Rapidminer does not require any computer programming knowledge as it provides a graphical user interface to design and execute analytical workflows (processes) that consist of multiple operators. Each of these performs a single task within the process, and the output of each operator forms the input of the next one.

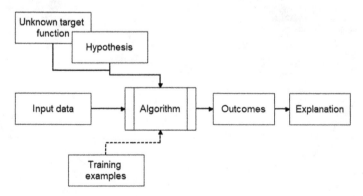

Figure 3.4 Schematic view of a machine learning model implementation

Table 3.4 Sample from the input data set

Places

Places
VENEZIA, GENOVA, VERONA, CREMONA
BRENNERO, CREMONA
PARMA, CREMONA, GENOVA, TORINO
MILANO, BRENNERO, CREMONA
BRESCELLO, MILANO, CREMONA
BRESCIA, VARESE, CREMONA
ROMA, CREMONA
TRENTO, CREMONA
ALESSANDRIA, BRENNERO, VERONA, CREMONA
CREMONA, TORINO
...

The input data set is a list of places visited by each respondent (Table 3.4). This is fed into the Rapidminer workflow (Figure 3.5). The workflow consists of four main steps:

(1) read the data set;
(2) clean the data and format in the required way;
(3) calculate the frequent itemsets (sets of locations that appear more frequently in many responses);
(4) derive the association rules.

The output contains a number of associations between different locations valued with the frequency with which they appear in the data set (support) and the probabilities (confidence) that a traveller visiting a certain

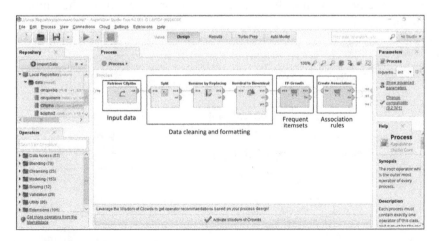

Figure 3.5 Rapidminer workflow

Table 3.5 Sample association rules for travellers coming to Cremona

From	To	Support	Confidence
MILANO	CREMONA	0.364	1.000
BRENNERO	CREMONA	0.091	1.000
BRESCIA	CREMONA	0.073	1.000
MILANO, BRENNERO	CREMONA	0.036	1.000
MILANO, NOVARA	CREMONA	0.036	1.000
BRENNERO, VERONA	CREMONA	0.036	1.000
VERONA, GENOVA	CREMONA	0.036	1.000
VERONA, VENEZIA	CREMONA	0.036	1.000
VERONA, GENOVA	CREMONA, VENEZIA	0.036	1.000
VERONA, VENEZIA	CREMONA, GENOVA	0.036	1.000
...			

location then goes to Cremona (Table 3.5), or that a Cremona tourist then visits another specific location (Table 3.6).

These outcomes can also be represented graphically, showing the most frequent locations for international tourists to Cremona (Figure 3.6) or departing from it (Figure 3.7).

This model can be particularly useful for those who want to target international tourists with specific promotions or campaigns. These actions can also be more effective if we combine the present results with those obtained previously on the main determinants of customer satisfaction.

Table 3.6 Sample association rules for travellers outgoing from Cremona

From	To	Support	Confidence
CREMONA	MILANO	0.364	0.364
CREMONA	BRENNERO	0.091	0.091
CREMONA	BRESCIA	0.073	0.073
CREMONA, VENEZIA	VERONA	0.036	1.000
CREMONA, VENEZIA	GENOVA	0.036	1.000
CREMONA, VERONA, GENOVA	VENEZIA	0.036	1.000
CREMONA, VENEZIA	VERONA, GENOVA	0.036	1.000
CREMONA, VERONA, VENEZIA	GENOVA	0.036	1.000
CREMONA, GENOVA, VENEZIA	VERONA	0.036	1.000
CREMONA, BRENNERO	MILANO	0.036	0.400
...			

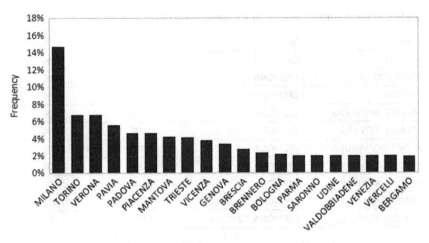

Figure 3.6 Most frequent origins of Cremona visitors

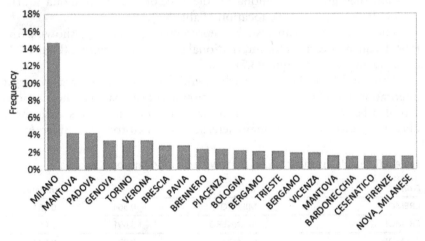

Figure 3.7 Most frequent destinations for Cremona visitors

Network Analysis

To study the structural characteristics of a destination, one of the most powerful tools available is the analysis of a network composed of the different components of a destination.

Any system, no matter how small or large, how simple, complicated or complex, human or artificial, is ultimately composed of several elements connected by some kind of relationship (Marmodoro & Yates,

2016). Therefore, it is difficult, if not impossible, to understand the nature and the behaviour of a system without fully grasping the 'network' behind it (Amaral & Ottino, 2004). When adopting a systemic view of a complex 'object' such as a tourism destination, it is natural to explore the methods made available by the so-called network science.

In a network study, the main objective is to count, map and analyse the patterns of relations between the various elements. These actors can be seen as the nodes or vertices of a network, connected by the relationships that exist between them (links or edges). The bases for these studies are rooted in the methods of the mathematical graph theory (Diestel, 2016), but they have undergone a number of improvements, modifications and adaptations and have delivered several metrics for measuring static and dynamic properties, and for producing models for the evolution of the systems considered (Barabási, 2016; Cimini *et al.*, 2019; da Fontoura Costa *et al.*, 2007).

In addition to providing an important framework for interdisciplinary approaches to tourism, network science has brought new insights into the study of tourism systems and has provided novel theoretic perspectives as well as tools for a better understanding that are useful for practitioners and policymakers (Baggio, 2017; Merinero-Rodríguez & Pulido-Fernández, 2016; van der Zee & Vanneste, 2015).

The network abstraction is rendered through a graph, a mathematical object made of nodes connected by edges, which can be represented by a matrix (adjacency matrix), thus allowing the use of linear algebra analysis methods (Figure 3.8).

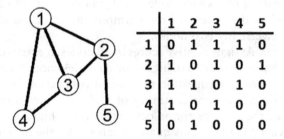

Figure 3.8 A simple network with its adjacency matrix

The inter- and multidisciplinary origin of network science, as previously discussed, has led to a wide variety of quantitative measurements of its topological (structural) characteristics. The literature on complex networks commonly uses the following measures to describe a network's structure:

- *Density*: The ratio between m and the maximum possible number of links that a graph may have.
- *Path*: A series of consecutive links connecting any two nodes in a network; the *distance* between two vertices is the length of the shortest path connecting them; the *diameter* of a graph is the longest distance (the maximum shortest path) existing between any two vertices in the graph; the *average path length* in the network is the arithmetic mean of all the distances. Numerical methods, such as the well-known Dijkstra algorithm, are used to calculate all the possible paths between any two nodes in a network.
- *Closeness*: The inverse of the sum of the distance from a node to all others, providing a measure of the degree to which an individual actor can reach all other network elements.
- *Clustering coefficient*: The degree of concentration of the connections of the node's neighbours in a graph, providing a measure of the local inhomogeneity of the link density. It is calculated as the ratio between the actual number of links connecting the neighbourhood (the nodes immediately connected to a chosen node) of a node and the maximum possible number of links in that neighbourhood. For the whole network, the clustering coefficient is the arithmetic mean single clustering coefficients.
- *Betweenness*: The number of shortest paths from all vertices to all others that pass through a node. It signals the importance of a node as a bridge between different network areas or as a bottleneck.
- *Efficiency* (at a global or local level): The ability of a networked system (global) or of a single node (local) to exchange information. It is calculated as the inverse of the harmonic mean of the shortest path length between any two nodes.
- *Eigenvector*: Assigns relative scores to all nodes in a network based on the idea that connections to high-scoring nodes contribute more to the score of the node in question than equal connections to low-scoring nodes. The eigenvector centrality of node n is the value of the nth element of the eigenvector associated with the highest eigenvalue of the adjacency matrix. This quantity can render the importance of an actor base not only on its own connectivity characteristics, but also on those of its neighbours.
- *Assortative mixing coefficient*: The Pearson correlation coefficient between the degree of a node and those of its first neighbours. If positive, the network is said to be assortative (otherwise disassortative). In an assortative network, well-connected elements (those with high degrees) tend to be linked to each other.

When non normalised by definition (clustering coefficient, for example), the individual metrics are usually normalised (in this case the term *centrality* is often used).

The matrix representation also allows the use of an algebraic approach to the analysis. For a square matrix, it is possible to calculate the eigenvalues and eigenvectors, characteristic quantities that summarise the topological properties of the matrix, and thus of the network (Restrepo *et al.*, 2006; Van Mieghem, 2010). More precisely, eigenvalues contain global information about the network, while eigenvectors contain local (nodal) information. The set of eigenvalues is called the *spectrum of the matrix*. This is the case, for example, for a number of nodal metrics such as eigenvector centrality previously mentioned (Bonacich, 1987) or PageRank (Brin & Page, 1998), all calculated from the principal (largest) eigenvector of the adjacency matrix. The spectral analysis of the adjacency and the Laplacian matrix of a network can be a useful and in many cases a computationally more efficient method to derive its main parameters. Among the many interesting conclusions of the wide body of studies in spectral graph theory, one concerns the connectivity characteristics and the possibility to use a spectrum to partition (hence to find communities), the second concerns the capacity of a network to control dynamic processes such as diffusion processes or the synchronisation of its members with respect to some features, opinions for example.

Among the different quantities, the distribution of the degrees of the nodes of a network is an important parameter of a network topology. This is usually expressed as a statistical probability distribution $N(k)$, i.e. for each degree present in the network, the fraction of nodes having that degree is calculated. The empirical distribution is then plotted and fit to find a functional relationship. Most real networks have been found to have a power-law degree distribution: $N(k) \sim k^{-\gamma}$. This is an important result because this type of relationship clearly signals the complex characteristics of the system.

The features of self-similarity and self-organisation, which are the most important characteristics of a complex system, are rendered (at least asymptotically) through a power-law distribution of certain parameters (size of components, number of connections, distribution of elements, etc.). Power laws are also associated with phase transitions or fractal features (Komulainen, 2004), and play a significant role in the description of those critical states between a chaotic and a completely ordered state, a condition known as *self-organised criticality* (Bak, 1996).

Moreover, such long-tailed distributions explain the typical resilience of a complex system that can, at the same time, be quite robust with respect to random shocks leading to the (undifferentiated) removal of nodes and have high fragility when targeted attacks are directed towards the most important (highly connected) elements (Barabási, 2016). In other words, finding a power law is confirmation of the complexity of networked systems.

A power–law relationship is scale invariant, i.e. no characteristic value can be defined to 'summarise' the parameter (in a Gaussian distribution this would be the arithmetic mean), and the behaviour of the parameter is the same when examined at different scales (Baggio, 2008), hence the term *scale-free* often associated with these systems.

A complex network may exhibit some form of substructure. Local groups can form that have denser within-group connections while having more sparse ties with nodes outside the group. The study of this modular structure of communities has attracted academic attention, since communities are a common trait of many real networked systems and may be central to understanding their organisation and evolution. Many methods exist for assessing these substructures. They rely on numerical algorithms that can identify some similarity in the local patterns of linking (Fortunato, 2010). In general, a measure called the *modularity index* is used to gauge the effectiveness of the outcomes. It represents a ratio between the density of the connections inside a module and the expected value of the same quantity that could be found in a graph of the same size but with a completely random distribution of links.

Combining the quantitative analysis with the essential qualitative knowledge of the object considered, a full model is built at three levels:

- *Macroscopic*: The global characteristics of the network. The most commonly used method to render a global topology is the statistical distribution of the degrees (degree distribution) that shows the general properties and its capacity to respond to dynamic processes. Other measures used to describe the macroscopic features are the average path length, the diameter, the density, the global efficiency in transferring information, the assortativity and the average values of the different metrics over the whole network.
- *Mesoscopic*: The intermediate structure, normally derived from a modularity analysis.
- *Microscopic*: The measures of the properties of the single elements (nodes). The most important quantities are the number of connections each node has (degree); the local density of links (clustering coefficient); or the extent to which a node is central for connecting different parts and thus acts as a bridge, or bottleneck (betweenness).

The analysis of a network starts with the collection of the data needed. This is a delicate task. Fully enumerating both nodes and links (mainly these) is usually quite difficult if not impossible. This is especially true for socioeconomic systems and is certainly the case for a tourism destination. Sampling these elements is possible but requires special care. Most of the quantities measurable in a network have statistical distributions that are very far from the well-known Gaussian shape. Standard statistical considerations on the significance of a sample apply only as long as we consider a system in which the elements are randomly placed, but this is usually not the case (see e.g. Cimini *et al.*, 2019; da Fontoura Costa *et al.*, 2011). The problem has been highlighted several times as a consequence of many network studies. For example, it has been found that in a structured network (scale-free, for example) it is not possible to easily determine the significance of a sample collected and that the topological characteristics of randomly extracted subnetworks may differ greatly from the structure of the whole system (Lee *et al.*, 2006; Stumpf *et al.*, 2005). Depending on the results of the analysis of the data available, the researcher needs to judge and make an educated guess on the final topology exhibited by the whole population; the whole network. In the cases in which this is possible, it can be found how some of the main network metrics vary with the size of the sample and the topology of the network. For example, according to the literature, in the case of an SF network, the degree distribution exponent and average path length decrease when nodes or links are sampled, there is little or no change to the assortativity coefficient and the clustering coefficient decreases when nodes are sampled, but increases when links are sampled (Kossinets, 2006).

A good data collection, therefore, uses a mixture of sources and techniques. The nodes of the network are the core tourism organisations (such as hotels, travel agencies, associations and public bodies), which can be easily identified from an official local tourism board. For the links, it is possible to start from a survey in which destination stakeholders are asked to specify their most important connections. These results are then complemented by publicly available documents or official records (co-ownership, membership in groups or consortia, interlocking directorates, etc., see Baggio [2018] and Christopoulos & Aubke [2014] for more details). The data obtained and their completeness can be further validated with a series of interviews with a selected sample of local knowledgeable informants (directors of the local tourism board and the main industrial associations, consultants active in the area, etc.).

The Cremona destination network was built following the lines previously sketched. The network comprises 90 elements whose distribution is detailed in Table 3.7.

Table 3.7 Cremona network composition

Business type	%
Associations	14
Travel agents	2
B&B	11
Hotel	36
Inn	8
Restaurants	18
Other services	11

The Cremona network and its degree of distribution are shown in Figure 3.9. This is compatible with a power-law $N(k) \sim k^{-\gamma}$ with $\gamma = 2.8 \pm 0.7$. The exponent is a signal of a preferential attachment mechanism for the formation of the network, a mechanism in which a new node connects with higher probability to a node that already has a high number of links (Barabási, 2016). The other main network measures are shown in Table 3.8.

The network looks rather sparse (low density) and relatively compact (low path length and diameter). High-degree nodes tend to connect to similarly well-connected nodes (assortativity), which confirms the formation mechanism previously inferred. The efficiency in transferring information along the network's connections is relatively low.

The clustering coefficient, as mentioned, measures the concentration of connections in the neighbourhood of a node and offers a measure of the heterogeneity of the local link density. A possible hierarchical

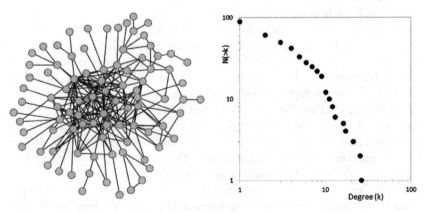

Figure 3.9 The Cremona network and its cumulative degree distribution

Table 3.8 Main Cremona network metrics

Metric	Value
Nodes	90
Links	223
Density	0.056
Average path length	3.315
Diameter	8
Global efficiency	0.358
Assortativity	0.011
Average clustering coefficient	0.26

organisation of the system (Ravasz & Barabási, 2003) is suggested by C and its distribution as a function of the nodal degrees, when the distribution of average clustering coefficients with respect to the degrees shows a power-law functional form: $C_{ave}(k) \sim k^{-\gamma}$. Figure 3.10 indicates that this holds, even if not very significantly.

A modularity analysis reveals a mesoscopic structure visible even if not very well defined (modularity index is $M = 0.442$), made of eight

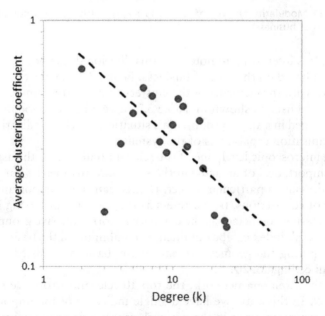

Figure 3.10 Average clustering coefficient as a function of degree (dotted line is a power–law relationship)

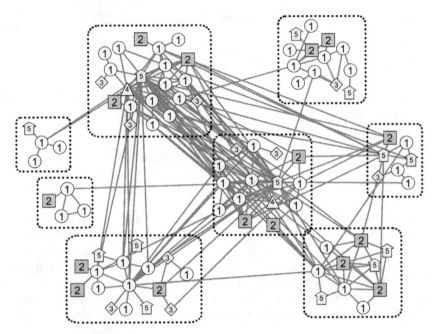

Figure 3.11 Modularity analysis of Cremona network (shapes and numbers distinguish types of business)

clusters. It is interesting to note how this division does not correspond to a division based on the type of business. Figure 3.11 shows the clustering obtained, with the elements of the clusters shaped (and numbered) using the type of business shown in Table 3.7 (here all accommodations have been arranged in a single group). This situation is a clear indication of the self-organisation capabilities of the system.

At a microscopic level, one of the relevant outcomes is the assessment of the 'importance' of an actor in the system. Many metrics can be used, each addressing a particular aspect. Degree can signal importance as the number of connections, betweenness as the role in transferring information, clustering coefficient as the capacity to form cohesive groups and so on. An overall index can be obtained by combining all the basic measures and computing the geometric mean of the basic normalised quantities (Sainaghi & Baggio, 2014).

For the Cremona network, the top 10 relevant actors are shown in Table 3.9. In this way, we have a single indicator of the importance of the position of an actor in the network representing the intensity of the activity in the community; if needed, this can be complemented by adding

Table 3.9 Most relevant actors in the Cremona network

Rank	Name	GeoMean
1	Commerce association	0.429
2	Po River consortium	0.404
3	Travel agency AV02	0.374
4	Hotel HO26	0.349
5	Hotel HO22	0.309
6	Hotel HO30	0.305
7	Inn LO02	0.302
8	Hotel HO34	0.295
9	Hotel HO17	0.287
10	Inn LO03	0.271

(averaging) some other features that may come from different investigations such as a survey on the relevance of those actors in the destination or some measure of intrinsic characteristics.

Agent-based Models

Within the possible techniques and methods to model a complex system, ABMs are effective tools when relationships, processes and elements cannot be easily (or at all) expressed in analytical terms. ABMs are computational models that simulate the actions and interactions of agents (Wilensky & Rand, 2015). Their importance essentially lies in their ability to expand the comprehension of phenomena and in providing a means for simulating behaviours and configurations.

These models have been used successfully in a number of different fields, from ecology to the social sciences to interacting coupled social–ecological systems, to study and simulate the properties, structures and dynamic processes of many natural and artificial systems (Baggio & Janssen, 2013; Bonabeau, 2002; Rand & Rust, 2011; Turci et al., 2015). They are especially suitable for dealing with social and economic environments, where general *laws* can seldomly be properly expressed in analytic terms (and most of the time cannot be solved), while interactions at the local level (between any two individual entities) can, instead, quite often be set with reasonable accuracy. Their applications to the tourism domain are relatively recent and not yet well diffused, but many outcomes are quite promising (Amelung et al., 2016; Johnson et al., 2017; Nicholls et al., 2017).

ABMs allow the simulation of a system from the bottom up, through an ensemble of individual entities called *agents*. The interactions between agents and between agents and the *environment* generate emergent static and dynamic configurations typically not possible to be implied just by *averaging* individual behaviours.

Building an ABM is a relatively straightforward task, provided a rigorous methodology is employed. The elements needed are a clear definition of the system or phenomenon to be modelled, a careful description of the different agents at play (with the features reputed essential for the purpose) and an accurate depiction of the interactions between them (Gilbert & Terna, 2000). Then the model can be coded and implemented.

Numerous programs, languages and platforms exist for realising an ABM (Abar *et al.*, 2017). One of the most used is NetLogo (Wilensky, 1999). It is an integrated modelling environment that includes a specific programming language and provides good facilities for building user interfaces. Its functional programming language allows even inexperienced users to produce an ABM with relative ease. Finally, NetLogo is designed to run on all common operating systems and an online web version has been made available that runs entirely in a browser.

Once a model has been coded, it can be executed and the outcomes explored. The modification of one or more of the parameters (agents' properties or interaction settings), after a number of cycles that imitate the temporal evolution of the system, will generate arrangements that can be examined, interpreted and used for a better understanding of the system or for formulating scenarios aimed at supporting policy and decision makers.

An important step for the production of a reliable and faithful model is the calibration of the parameters used. This activity consists of finding a set of values, normally based on empirical observations or on the literature, that can reproduce known outcomes. In other words, the model needs to be trained (as an ML algorithm) and verified. For example, one could devise a model to assess how different behavioural schemes lead to specific tourist flows in a certain geographic area. In this case, it would be possible to calibrate the initial parameters trying to reproduce observed quantities.

A second important step is a sensitivity analysis, which is the analysis of how and to what extent changes in individual parameters may influence the outcomes (ten Broeke *et al.*, 2016).

The population of agents used in an ABM can be distributed in several ways, depending on the system studied and the objectives of the model. Given the importance of the relationships between elements assumed in such systems, it is natural to use the results of a network analysis as a substrate for modelling structural or dynamic processes.

As an example, we design an ABM for investigating the diffusion of information in the Cremona destination. For the mechanism, we use the analogy with the diffusion of a disease (Hethcote, 2000).

The simplest model is based on the cycle of infection in a population. The N individuals may be in one of two possible states: susceptible (S) to infection or infected (I), when they contact the disease. In our case, the disease is a piece of information so individuals can possess the information (*infected*) and are able to transmit it to those who do not have the information (*susceptible*), informing (*infecting*) them if they are connected in some way. We use as a basis the network obtained previously. The agents, thus, are the stakeholders identified. As a feature, we define an actor's capacity to receive and transfer information that we derive from the importance index calculated, mediated with the actor's intrinsic characteristics (organisational efficiency, capacity to handle new information, etc.). For simplicity, we split the set of values into three levels: high, medium and low. To each we assign a probability to transfer information to its neighbours that is one of the main parameters of the model. In other words: one agent is infected and can transfer a piece of information to its neighbours with a probability that depends on the importance index. It is also possible to set a fraction of neighbours that can be *infected* at each time step.

The model is realised in NetLogo (Figure 3.12). We perform two simulations: one using the different transmission probabilities and the second one disregarding these differences. The model is aimed at understanding the results of some actions directed towards an elimination (or a sensible decrease) in these capacities that could be obtained, for example, with some educational programme or with some plan for improving the efficiency of an organisation and its capacity to communicate with others.

The results of the two simulations are shown in Figure 3.13, which reports the cumulative fraction of infected agents and their fraction in the different time periods (differential distribution). Given the stochastic nature of the model, these are the averages of 10 runs.

The effects of the choices made and of the structural characteristics of the system are clear. The best possible results in terms of speed and the extent of diffusion can be obtained when the capacities of the actors are, as much as possible, at the same level, both in terms of their internal abilities and their connectivity, almost independently from what the level is (this can be possibly tuned by changing the probabilities of transmission). That is to say, if we provide the largest number of stakeholders with appropriate capacities and organisational efficiency, maybe supported by the right toolset, we obtain the best possible results. This can help in improving plans and policies.

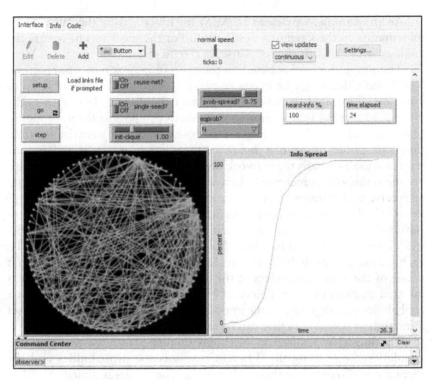

Figure 3.12 The NetLogo diffusion model

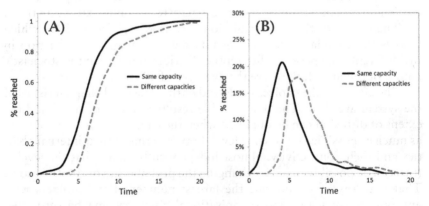

Figure 3.13 Simulation results: (A) the cumulative distribution of infected individuals and (B) the differential distribution

System Dynamic Models

A complex system is usually a dynamic entity, and SD modelling is a useful computer-based approach for a good understanding of a system's behaviour over time.

SD allows modelling of complex systems by considering their major components and defining the connections between them as causal relationships. In essence, in a system dynamic model, we define interactions in terms of feedback processes and flows of resources between the different components (termed *stocks*), possibly having time delays between causes and effects (Sterman, 2001).

Also in this case, we start from an initial CM in which system elements and relations are specified as feedback loops. These can be positive or negative. Positive feedbacks (self-reinforcing) reinforce the process, amplifying the effects on the state of the system. This amplification may lead to exponential unlimited growth, so positive feedbacks are balanced by negative (limiting) effects that tend to balance the overall dynamics.

The SD approach combines qualitative and quantitative aspects. The starting point is usually a causal loop diagram (CLD) in which a qualitative description of the structure, the main elements and the feedback relationships are graphically rendered (Figure 3.14).

The quantitative features, often expressed as differential equations, are used for building a *stock and flow* diagram (Figure 3.15). Here, the system is presented in terms of *stocks*, variables expressing some quantity existing at a given time (that might also have accumulated in the past) and *flows*, dynamically evolving variables that can contribute to increasing the stock (in-flows) or diminishing its value (out-flows).

As an example, let us consider a process of diffusion. It can be modelled using the proposal of Bass (1969). Originally designed to explain the adoption of a new product, the model has been used in a wide variety of settings and is able to model processes such as the adoption of a new technology or an innovation, or the reception of a promotional message.

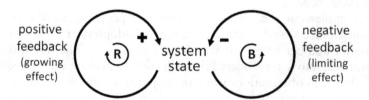

Figure 3.14 Causal loop diagram (R is the reinforcing side, B is the balancing side)

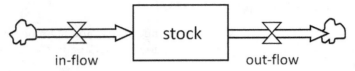

Figure 3.15 Stock and flow diagram

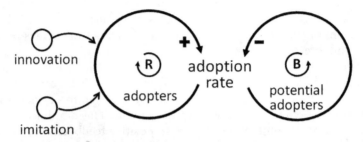

Figure 3.16 CLD for a Bass diffusion model

In its most basic version, the model consists of a simple differential equation that describes how new products are adopted by a population. The adopters are classified as innovators or imitators and the speed and timing of an adoption depends on two factors: one is connected to the innovative characteristics of the new product and the second depends on external factors such as government policies, mass media communications and competition. Here, the limiting factor is obviously the decreasing number of adopters, since the starting population, no matter how large, is of a finite size.

A CLD is shown in Figure 3.16 and the corresponding stock and flow diagram is shown in Figure 3.17.

Several software applications exist for designing and running an SD model. The most used are listed on the Wikipedia page: en.wikipedia.org /wiki/Comparison_of_system_dynamics_software. Many of them are proprietary commercial applications, but facilities exist for academic or educational uses.

To implement the model, we need to operationalise the relevant variables. Since we will output a fraction of adopters at each time step, the initial number of potential users is not critical. The imitation and innovation factors are rendered through coefficients that can be deducted from empirical observations when the object of study (the product or the technology etc.) has already started its life. If these are unknown, the Bass model allows analogy to be used in determining the coefficients (Baggio

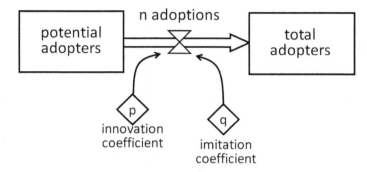

Figure 3.17 Stock and flow diagram for the Bass model

& Caporarello, 2003; Mahajan *et al.*, 1986). An interesting distinctive feature of the Bass model, with its many extensions and modifications, is its capacity to forecast future trends based on an assessment of the current situation (Mahajan *et al.*, 1995).

SD models do not take into account the microscopic structure of the system studies. Therefore, the possible effects of the arrangement of the different relationships, when significant, must be reflected in the variables used. In our case, knowing that the topology of the network formed by the Cremona stakeholders can greatly affect a diffusion process, we can act on the imitation coefficient q using values that favour faster spreads.

A typical outcome from a run of the diffusion model sketched so far is a curve showing the time evolution of the cumulative number of adopters. By changing the two parameters (p and q), it is possible to gauge the

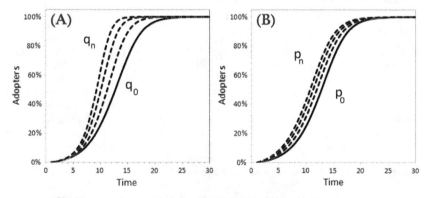

Figure 3.18 Bass model outcomes when changing simulation parameters (A: changing q values; B: changing p values; solid line is baseline)

effects they have on the overall process. This can be a useful indication for setting policies or designing strategies. An example is shown in Figure 3.18 (Panel A: $q_0 < \cdots < q_n$; Panel B: $p_0 < \cdots < p_n$). The higher-value curves (dotted lines in Figure 3.18) show both the speed up and the wider extent of the diffusion process. It is clear, however, that the variation (increase) in the imitation parameter has a much greater effect than the increase in the innovation parameter.

Interpreting these outcomes in our context, we can identify the imitation parameter as resulting from the internal communications in the destination (e.g. word of mouth) and the innovation parameter as a kind of 'advertising' effort made by some organisation (e.g. a destination marketing organisation or a tourism board). The indications provided by the different scenarios can thus better inform the actions that should be taken.

Concluding Remarks

In this chapter, we have described some of the most important modelling techniques that can be employed for answering the questions we want or need to assess. The different methods have been accompanied by worked examples.

The objective of this roundup has been to show some real cases of interest. Using the same system, the Cremona destination, has helped highlight the different possible uses of the methods discussed and may enable the design of a more complex, but complete, approach to the issue of understanding both the phenomenon and the system itself. These two aspects, as discussed in Chapter 1, cannot be easily separated when a complex system is the object of study.

References

Abar, S., Theodoropoulos, G.K., Lemarinier, P. and O'Hare, G.M. (2017) Agent-based modelling and simulation tools: A review of the state-of-art software. *Computer Science Review* 24, 13–33. doi:10.1016/j.cosrev.2017.1003.1001

Amaral, L.A.N. and Ottino, J.M. (2004) Complex networks: Augmenting the framework for the study of complex systems. *The European Physical Journal B* 38, 147–162.

Amelung, B., Student, J., Nicholls, S., Lamers, M., Baggio, R., Boavida-Portugal, I., Johnson, P., de Jong, E., Hofstede, G.-J., Pons, M., Steiger, R. and Balbi, S. (2016) The value of agent-based modelling for assessing tourism-environment interactions in the Anthropocene. *Current Opinion in Environmental Sustainability* 23, 46–53.

Baggio, R. (2008) Symptoms of complexity in a tourism system. *Tourism Analysis* 13 (1), 1–20.

Baggio, R. (2017) Network science and tourism: The state of the art. *Tourism Review* 72 (1), 120–131.

Baggio, R. (2018) Network analysis: Quantitative methods in tourism. In C. Cooper, B. Gartner, N. Scott and S. Volo (eds) *The Sage Handbook of Tourism Management* (pp. 150–170). Thousand Oaks, CA: SAGE.

Baggio, R. and Caporarello, L. (2003) *Gestire la tecnologia: metodi di analisi e valutazione.* Milano: Guerini.

Baggio, J.A. and Janssen, M.A. (2013) Comparing agent-based models on experimental data of irrigation games. In R. Pasupathy, S.-H. Kim, A. Tolk, R. Hill and M.E. Kuhl (eds) *Proceedings of the 2013 Winter Simulation Conference* (pp. 1742–1753). Piscataway, NJ: IEEE Press.

Baggio, R. and Klobas, J. (2017) *Quantitative Methods in Tourism: A Handbook* (2nd edn). Bristol: Channel View Publications.

Bak, P. (1996) *How Nature Works. The Science of Self-Organized Criticality.* New York: Springer.

Barabási, A.L. (2016) *Network Science.* Cambridge: Cambridge University Press.

Bass, F.M. (1969) A new product growth model for consumer durables. *Management Science* 15 (5), 215–227.

Bishop, C.M. (2006) *Pattern Recognition and Machine Learning.* New York: Springer-Verlag.

Bonabeau, E. (2002) Agent-based modeling: Methods and techniques for simulating human systems. *Proceedings of the National Academy of the Sciences of the USA* 99, 7280–7287.

Bonacich, P. (1987) Power and centrality: A family of measures. *American Journal of Sociology* 92, 1170–1182.

Brin, S. and Page, L. (1982) The anatomy of a large-scale hypertextual (web) search engine. *Computer Networks and ISDN Systems* 30 (1–7), 107–117.

Christopoulos, D. and Aubke, F. (2014) Data collection for social network analysis in tourism research. In M. McLeod and R. Vaughan (eds) *Knowledge Networks and Tourism* (pp. 126–142). New York: Routledge.

Cimini, G., Squartini, T., Saracco, F., Garlaschelli, D., Gabrielli, A. and Caldarelli, G. (2019) The statistical physics of real-world networks. *Nature Reviews Physics* 1 (1), 58–71.

Cloutier, R.J. and BKCASE Editorial Board (eds) (2017) *The Guide to the Systems Engineering Body of Knowledge (SEBoK), v. 1.9.1.* Hoboken, NJ: The Trustees of the Stevens Institute of Technology.

da Fontoura Costa, L., Rodrigues, A., Travieso, G. and Villas Boas, P.R. (2007) Characterization of complex networks: A survey of measurements. *Advances in Physics* 56 (1), 167–242.

da Fontoura Costa, L., Oliveira, O.N., Travieso, G., Rodrigues, F.A., Villas Boas, P.R., Antiqueira, L., Viana, M.P. and Correa Rocha, L.E. (2011) Analyzing and modeling real-world phenomena with complex networks: A survey of applications. *Advances in Physics* 60 (3), 329–412.

Diestel, R. (2016) *Graph Theory* (5th edn). New York: Springer. See http://diestel-graph-theory.com/ (accessed December 2016).

Fan, J., Han, F. and Liu, H. (2014) Challenges of Big Data analysis. *National Science Review* 1 (2), 293–314.

Fortunato, S. (2010) Community detection in graphs. *Physics Reports* 486 (3–5), 75–174.

Franks, B. (2012) *Taming the Big Data Tidal Wave: Finding Opportunities in Huge Data Streams with Advanced Analytics.* Hoboken, NJ: John Wiley and Sons.

Gallarza, M.G., Saura, I.G. and García, H.C. (2002) Destination image: Towards a conceptual framework of tourism research. *Annals of Tourism Research* 29 (1), 56–78.

Gandomi, A. and Haider, M. (2015) Beyond the hype: Big data concepts, methods, and analytics. *International Journal of Information Management* 35 (2), 137–144.

García, S., Luengo, J. and Herrera, F. (2015) *Data Preprocessing in Data Mining*. New York: Springer.

Gilbert, N. and Terna, P. (2000) How to build and use agent-based models in social science. *Mind & Society* 1 (1), 52–72.

Granville, V. (2013) The Curse of Big Data. See http://www.analyticbridge.com/profiles/b logs/the-curse-of-big-data (accessed June 2014).

Hethcote, H.W. (2000) The mathematics of infectious diseases. *SIAM Review* 42 (4), 599–653.

Johnson, P., Nicholls, S., Student, J., Amelung, B., Baggio, R., Balbi, S., Boavida-Portugal, I., de Jong, E., Hofstede, G.-J., Lamers, M., Pons, M. and Steiger, R. (2017) Easing the adoption of agent-based modelling (ABM) in tourism research. *Current Issues in Tourism* 20 (8), 801–808.

Kao, M.-Y. (ed.) (2016) *Encyclopedia of Algorithms* (2nd edn). Berlin: Springer.

Keogh, E. and Mueen, A. (2011) Curse of dimensionality. In C. Sammut and G.I. Webb (eds) *Encyclopedia of Machine Learning* (pp. 257–258). New York: Springer.

Kim, Y.G., Eves, A. and Scarles, C. (2009) Building a model of local food consumption on trips and holidays: A grounded theory approach. *International Journal of Hospitality Management* 28 (3), 423–431.

Kim, Y.G., Eves, A. and Scarles, C. (2013) Empirical verification of a conceptual model of local food consumption at a tourist destination. *International Journal of Hospitality Management* 33, 484–489.

Komulainen, T. (2004) Self-similarity and power laws. In H. Hyötyniemi (ed.) *Complex Systems: Science on the Edge of Chaos (Report 145, October 2004)*. Helsinki: Helsinki University of Technology, Control Engineering Laboratory.

Konishi, S. and Kitagawa, G. (2008) *Information Criteria and Statistical Modeling*. New York: Springer.

Kossinets, G. (2006) Effects of missing data in social networks. *Social Networks* 28 (3), 247–268.

Lee, S.H., Kim, P.-J. and Jeong, H. (2006) Statistical properties of sampled networks. *Physical Review E* 73, 102–109.

Leung, D., Law, R., van Hoof, H. and Buhalis, D. (2013) Social media in tourism and hospitality: A literature review. *Journal of Travel & Tourism Marketing* 30 (1–2), 3–22.

Mahajan, V., Mason, C.H. and Srinivasan, V.S. (1986) An evaluation of estimation procedures for new product diffusion models. In V. Mahajan and Y. Wind (eds) *Innovation Diffusion Models of New Product Acceptance* (pp. 203–232). Cambridge, MA: Ballinger.

Mahajan, V., Muller, E. and Bass, F.M. (1995) Diffusion of new products: Empirical generalizations and managerial uses. *Marketing Science* 14, G79–G88.

Marmodoro, A. and Yates, D. (eds) (2016) *The Metaphysics of Relations*. Oxford: Oxford University Press.

Mayer-Schönberger, V. and Cukier, K. (2013) *Big Data: A Revolution that will Transform How We Live, Work, and Think*. New York: Houghton Mifflin Harcourt.

Merinero-Rodríguez, R. and Pulido-Fernández, J.I. (2016) Analysing relationships in tourism: A review. *Tourism Management* 54, 122–135.

Mierswa, I. and Klinkenberg, R. (2018) RapidMiner Studio, Boston, MA. See www.rapidminer.com (accessed December 2018).

Milo, P. (2015) Destination management e il Capitale Sociale: il caso della provincia di Cremona. Unpublished MSc thesis, Libera Università di Lingue e Comunicazione IULM, Milan.

Mitchell, T.M. (1997) *Machine Learning*. New York: McGraw Hill.

Nicholls, S., Amelung, B. and Student, J. (2017) Agent-based modeling: A powerful tool for tourism researchers. *Journal of Travel Research 56* (1), 3–15.

Rand, W. and Rust, R.T. (2011) Agent-based modeling in marketing: Guidelines for rigor. *International Journal of Research in Marketing 28* (3), 181–193.

Ravasz, E. and Barabási, A.-L. (2003) Hierarchical organization in complex networks. *Physical Review E 67* (2), 026112.

Restrepo, J.G., Ott, E. and Hunt, B.R. (2006) Characterizing the dynamical importance of network nodes and links. *Physical Review Letters 97*, art. 094102.

Robinson, S. (2008a) Conceptual modelling for simulation Part I: Definition and requirements. *Journal of the Operational Research Society 59* (3), 278–290.

Robinson, S. (2008b) Conceptual modelling for simulation Part II: A framework for conceptual modelling. *Journal of the Operational Research Society 59* (3), 291–304.

Sainaghi, R. and Baggio, R. (2014) Structural social capital and hotel performance: Is there a link? *International Journal of Hospitality Management 37*, 99–110.

Sammut, C. and Webb, G.I. (2017) *Encyclopedia of Machine Learning and Data Mining* (2nd edn). Berlin: Springer.

Samuel, A. (1959) Some studies in machine learning using the game of checkers. *IBM Journal 3*, 211–229.

Smith, S.L.J. (2013) *Tourism Analysis: A Handbook* (2nd edn). London: Routledge.

Sterman, J.D. (2001) System dynamics modeling: Tools for learning in a complex world. *California Management Review 43* (4), 8–25.

Stumpf, M.P.H., Wiuf, C. and May, R.M. (2005) Subnets of scale-free networks are not scale-free: Sampling properties of networks. *Proceedings of the National Academy of the Sciences of the USA 102* (12), 4221–4224.

ten Broeke, G., van Voorn, G. and Ligtenberg, A. (2016) Which sensitivity analysis method should I use for my agent-based model? *Journal of Artificial Societies and Social Simulation 19* (1), art. 5.

Turci, L., Pennec, S., Toulemon, L., Bringé, A., Baggio, R. and Morand, E. (2015) Agent-based microsimulation of population dynamics. In M. Bierlaire, A. de Palma, R. Hurtubia and P. Waddell (eds) *Integrated Transport & Land Use Modeling for Sustainable Cities* (pp. 113–135). Lausanne: EPFL Press.

van der Zee, E. and Vanneste, D. (2015) Tourism networks unravelled: A review of the literature on networks in tourism management studies. *Tourism Management Perspectives 15*, 46–56.

Van Mieghem, P. (2010) *Graph Spectra for Complex Networks*. Cambridge: Cambridge University Press.

Veal, A.J. (2006) *Research Methods for Leisure and Tourism: A Practical Guide* (3rd edn). Harlow: Financial Times/Prentice Hall/Pearson Education.

Wang, C., Chen, M.H., Schifano, E., Wu, J. and Yan, J. (2016) Statistical methods and computing for big data. *Statistics and Its Interface 9* (4), 399–414.

Wilensky, U. (1999) NetLogo. http://ccl.northwestern.edu/netlogo. Evanston, IL: Center for Connected Learning and Computer-Based Modeling. Northwestern University.

Wilensky, U. and Rand, W. (2015) *An Introduction to Agent-Based Modeling: Modeling Natural, Social, and Engineered Complex Systems with NetLogo*. Cambridge, MA: MIT Press.

Witten, I.H., Frank, E., Hall, M.A. and Pal, C.J. (2016) *Data Mining: Practical Machine Learning Tools and Techniques*. Cambridge, MA: Morgan Kauffman.

Zikopoulos, P.C., Eaton, C., DeRoos, D., Deutsch, T. and Lapis, G. (2012) *Understanding Big Data*. New York: McGraw-Hill.

4 Advanced Modelling Methods

Rapid advances in the methods and techniques used for modelling and simulating have been driven by research on numerical algorithms and the wide availability of data provided by a profusion of online applications and platforms. Even the tourism domain, albeit at a slower pace, is recognising the role of these practices and is increasingly using the results of this large interdisciplinary effort. In this chapter, we describe and discuss two main areas where progress has been made in the last few years, particularly network analysis and artificial intelligence.

Tourism, as we have seen, is a complex adaptive system, composed of multiple entities that interact at different scales and in different ways. As such, while simplification is helpful to better analyse and understand a system, too much simplification may hinder important effects mainly in the assessment of the relationships that exist regarding interactions and outcomes. Many variations of the different models discussed Chapter 3 can be employed. Of these, two refer to the use of network science. As mentioned, there has been significant development in this area in recent years, but few of the advanced techniques devised are commonly used in the tourism domain. Two of the more advanced methods have shown great potential and have been utilised to analyse systems when different levels of organisation exist or when multiple types of relationships exist between the system's components: exponential random graph modelling (Robins *et al.*, 2007; Wang *et al.*, 2009) and multiplex or multilayer network analysis (De Domenico *et al.*, 2013; Kivelä *et al.*, 2014).

Exponential random graph modelling allows one to inspect a network based on micro-level configurations (also called motifs) and provides methods to assess the significance of a structure's distributions. That is, the type of micro-level configuration prevalent (or not prevalent) in a specific tourism network. A multiplex/multilayer analysis allows for a more macro-level analysis of a network in which multiple relationships exist. This method considers the interplay of several layers composed of different actors connected by the same relationships, or the same agents whose connections are of a different origin, so that only some coarse (and often even illogical) simplification can lead to consider them as similar in nature.

Exponential Random Graphs Models

Exponential random graph models (ERGMs) deal with the overall network structure, considering it a product of micro-level patterns (motifs). In other words, ERGMs assume that the observed network is the result of some local process represented by specific configurations. Edges in the network are the result of a conditional probabilistic outcome (i.e. their probability depends on specific assumptions made by the researcher and by the other edges present in the network), hence the network is not the result of a random formation process (Wang et al., 2009).

An ERGM is, in essence, a statistical model that allows regularities in the network to be captured while recognising potential uncertainties, similar to any other statistical model. These allow inferences about whether certain properties or attributes are observed more or less frequently than that produced by pure chance, and based on these considerations derive clues on the importance (significance) of these and their effects. In exponential random graph modelling, motifs (or parameters) are enumerated and an assessment is made on whether they are observed more or less than expected by chance.

For other mathematical or statistical techniques, an ERGM, when correctly interpreted, can help develop hypotheses about the social processes that might have generated or facilitated the relationships observed in a network. Further, ERGMs allow testing of specific hypotheses about the structural features of a system, such as whether homogeneous groups are more likely to be linked by chance, or actors' relevance is more likely to be due to a random modification rather than having a 'real' value. For example, a typical question could be: do accommodations preferably link (i.e. interact, share knowledge and information, have dealings with) with other accommodations or stakeholder that perform diverse or complementary services such as restaurants and bars?

In a tourism context, ERGMs can also be used to assess whether specific forms of collaboration are more effective than others: are collaborations that are mandated by law (imposed in some way) more or less effective than collaborations that are born spontaneously from shared interests? Are country-level organisations more likely to collaborate with local-level organisations compared to other country-level organisations or vice versa? Some of these issues, as seen in Chapter 3 on network analysis, can be found using a basic network analysis. What ERGMs add to these analyses is an assessment of their effect size, thereby giving more complete and trustworthy answers to the questions posed.

While ERGMs were initially developed to analyse undirected, unweighted networks, lately they have been expanded to analyse networks composed of different types of nodes: multilevel ERGMs (Lazega & Snijders, 2016; Wang *et al.*, 2013). Multilevel ERGMs can, in principle, shed light on the linkages and characteristics of the networks of nested organisations. For example, individual hotels may share information, but they may also be part of a hotel chain that may or may not share information and may or may not hinder or facilitate collaboration at the local level.

In short, an ERGM allows one to understand the extent to which a complex social phenomenon is the result of both regularities or interdependencies and 'randomness', to make deductions on whether certain network signatures appear more often than would be expected by chance alone. Moreover, these methods are useful in distinguishing between different social processes that may have similar consequences and in better understanding how local social processes interact and combine to shape global network patterns. Thus, it is possible to arrange and quantify exogenous and endogenous effects and simultaneously model the effects of network structures, actor attributes and relational attributes, obtaining estimates for and the significance of each effect. In a simulation environment, it is also possible to exploit a further characteristic of ERGMs: they are generative models. That is, given a set of statistics on network structures and variables of interest, it is possible to generate networks that are consistent with any set of parameters.

The basic idea is that any of the many metrics that describe the structural features of a specific network (e.g. density, centrality and clustering) can only apply to the observed network, which is only one occurrence of many possible alternatives. This set of alternative objects may have similar or completely different features. A statistical model should take into account the set of all possible alternatives weighted on their similarity to the observed network. However, since network data are intrinsically relational, the assumptions of independence and a uniform distribution of standard models such as regressions, for example, are violated.

An ERGM works by computing a limited set of known statistics from the network and using distributions such as statistics to generate a family of randomly generated isomorph networks, which are usually the result

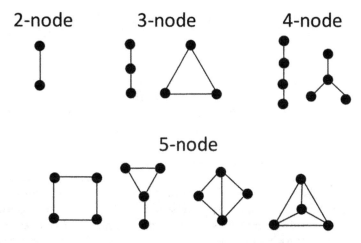

Figure 4.1 Graphlets examples (adapted from Pržulj, 2007)

of an enumeration of basic micro-configurations (motifs or graphlets, see Figure 4.1).

Once the distribution of these motifs in the network has been obtained, a number of randomly generated networks having the same motifs distributions are used for assessing the likelihood of the original statistics. The method uses an exponential family theory for specifying the probability distribution for a set of random graphs. The result is a maximum likelihood estimate for the parameters and a goodness of fit. It is possible to perform various comparisons and simulate other networks with the underlying probability distribution specified by the model.

ERGMs, as previously noted, can be a powerful tool to analyse social-ecological networks. However, a researcher needs to be careful when deciding to analyse a network in such a way. The following are the main steps to follow in building a reliable model; all steps should be completed, even if only the last step is of interest (parameter estimation and significance; see Robins *et al.* [2007]):

(1) Start by fully defining the network.
(2) Once defined, each edge in an ERGM is considered a random variable. The ERGM is not deterministic, thus its results are probabilistic (similarly to other simulations and computational models).
(3) Then variables or structures of interest are assessed. They can be global or local (nodal) properties or motifs configurations, which are the variables of the model. It is assumed that the probability of

each variable having a certain value with respect to a random assignment is conditional on all the other variables (i.e. variables are not independent).

(4) Finally, the model is estimated and the outcomes interpreted.

Obviously, we need be aware of the fact that if the model is complex, it may not give any outcome. In other words, the optimisation function may be unable to reach a solution and hence fail to provide an interpretable output.

As a first example, we consider an analysis of interorganisational cooperation in the domain of sport tourism (Wäsche, 2015). In this work, the relational structures that lie behind the sport tourism programme of two neighbouring destinations were studied. The actors contributing to the sport tourism programmes were identified and the network configuration analysed. An ERGM was built to test the significance of particular patterns in the formation of the network.

Three significant structural effects were highlighted: a positive homophily effect for organisations with the same aim in terms of a for-profit or not-for-profit orientation; multiple triangulation and multiple connectivity configurations; and the relevance of a private sport agency as the most central actor in the network.

Patterns of tourists' flows across countries is an important topic and greatly influences the whole phenomenon, an understanding of which is crucial for all operational and strategic activities in the field. It is obvious then that all peculiarities of these patterns need to be well studied. Such study has been conducted, for instance, by Lozano and Gutiérrez (2018). The authors began by collecting the World Tourism Organisation data on the inbound and outbound movements of international tourists between countries. They assembled a scale-free network with a high degree of centralisation, a sizeable clustered structure and a pronounced geographic homophily, that is: based on node attributes, similar nodes are more likely to connect to each other and form more densely inter-linked groups. This structure arises from geographical, trade and cultural proximity, where the main global or regional powers seem to have their own tourism sphere of influence.

The main motifs that occurred with high frequency were identified. These motifs included transitive feedforward loops (i.e. 'triangle' link configurations of the type: A to B to C to A) and different mutual dyad subgraphs (i.e. link configurations of the type: A to B to A). An ERGM inquiry confirmed the significance of these local substructures, explaining the observed topology and suggesting that some underlying mechanisms must be in play to produce them. In this case, it is suggested that the global mobility patterns observed can be explained as emerging from the superposition of local effects. In any region, intraregional travel is the main source of outbound and inbound tourism. The study also identified differences in the macro-regional patterns, showing that although the main destinations for most European countries are other European countries, a similar effect is less noticeable in Asia or North America. On the other hand, when inbound travel was considered, Asian countries have other Asian countries as source markets, while such configuration is less noticeable in Europe or North America, whose main source markets are more differentiated.

Although these results might seem relatively known or expected, the use of a rigorous methodology gives them more strength and reliability. Moreover, the technique can also be extended and used on a smaller geographical scale, thereby providing regional destination managers with better tools for understanding tourists' movements in their areas and allowing better grounded strategies and actions to be set.

Multilayer and Multiplex Networks

While exponential random graph modelling allows a researcher to understand the configuration of macro-level structures, albeit probabilistically, and the results of micro-level interactions and, by extension, how such micro-level interactions may affect specific outcomes of interest, the use of multilayer networks allows the researcher to look at complex macro-level network properties and local nodal properties in complex, interacting networks of networks.

In many systems, the interactions of the different entities can be more complicated than those modelled by a 'simple' network abstraction, because they can encompass multiple types of relationships, changes in time or multiple types of actors. Such systems include multiple subsystems and may have multiple layers of connectivity. It is thus important to consider these multilayer features to try to improve our understanding of the systems. Therefore, we might need to generalise network theory by developing a framework and tools to study such configurations in a comprehensive way.

In multilayer networks, not only is each layer a network in itself, but it is also part of a wider networked system. Each layer is defined, as for simpler classic networks (i.e. also termed *monoplex* networks), by the nodes and the relationship between them.

Here, however, there can be multiples nodes and relationships between them and the different layers can be connected in multiple ways.

A crucial aspect that controls the collective behaviour of a multilayer network is the way in which the interaction between the different types of connections is established. In particular, interlayer interactions have been shown to account for the emergence of novel phenomena due to the structural and dynamic correlations between the components of the system.

Figure 4.2 shows an example of a multilayer network in which the main elements are identified: nodes, interlayer links (edges that connect nodes between different layers) and intra-layer links (edges that connect nodes within the same layer).

Multilayer networks, in which the same set of nodes exists in every layer, are called *multiplex networks*. In multiplex networks, however, the nodes in each layer can be linked by different types of relationships (Aleta & Moreno, 2019).

In the tourism domain, for example, a multilayer network can be composed of the same stakeholders in each layer but connected in different ways: one layer could include economic or business relationships, another could consider information or opinion exchanges. This would allow, for example, to model the mutual effects of information interactions on the structure and the dynamics of the economic transactions performed in the system.

A multilayer network analysis allows for a more in-depth understanding of the structural properties of a complex interdependent network, avoiding unnecessary aggregation that reduces information (De Domenico *et al.*, 2013; Kivelä *et al.*, 2014).

A naturally multilayered network could be studied with the 'simpler' models discussed in Chapter 3 by aggregating the different types

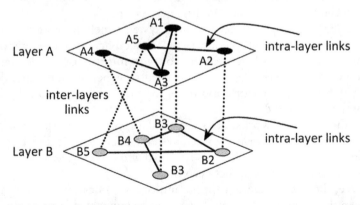

Figure 4.2 Multilayer network

of linkages and 'flattening' the network to a single layer. However, the aggregation process can neglect a lot of valuable information about the system and its behaviours. Developing models for multilayer networks is useful for improving our knowledge of the structure and functions of multilayer systems and can lead to discoveries of new phenomena that cannot be explained using a monoplex network approach. For example, the international trade network, an aggregated network that uses the total value of trade between countries as edge weights, is unable to capture the richness of the structures in the multiplex network in which each layer concerns different categories of products (Kivelä *et al.*, 2014). Aggregation can distort the properties of multiplex networks, mainly when non-trivial interlayer connections exist. In such cases, information related to connections between layers (i.e. the interlayer links) disappears when an aggregation process is used.

It must be noted here that given the amount of information on the key elements that a multilayer network requires, the time and cost for collecting data suitable for a multilayer network analysis are higher than in the case of a simplified aggregated network analysis. A generalised modelling technique allows us not only to improve our understanding of these specific configurations, but also to develop a unified framework covering all possible types of interdependence between networked elements.

Multilayer network representation

As seen in Chapter 3 on network analysis, a network can be represented via an adjacency matrix. By extension, multilayer networks can be represented by a tensor, a multidimensional array, in which the different 'dimensions' correspond to the layers and the interlayer connections. This is a generalisation of the algebraic approach used in traditional network analysis; in mathematical terms, the bidimensional adjacency matrix is a rank-2 tensor.

Technically, the tensor representation is only appropriate for a multilayer network that is node aligned, that is it has the same set of nodes in all layers. Although it is possible to insert dummy nodes to fill up the less numerous layers, this is a delicate task, as it can modify the overall structure. One solution is to adopt a different matrix representation, where the tensor is 'flattened' to a bidimensional matrix called a *supra-adjacency matrix*. This is a block matrix formed by assembling the single layers' adjacency matrices on the main diagonal and completing it with the matrices that contain the interlayer links. A bidimensional matrix is a much more familiar object and is more convenient when writing software

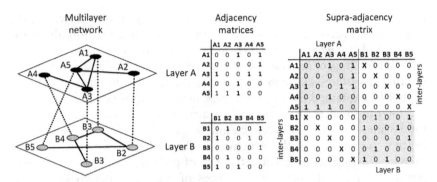

Figure 4.3 Multilayer network with its supra-adjacency matrix representation

implementations of the algorithms for computing the different matrices. Figure 4.3 is a graphical representation of a multilayer network, of the layers' adjacency matrices and of the overall supra-adjacency matrix.

The relationship between classic network metrics and multilayer metrics is slightly more complicated, but their meaning is the same, and their extension is, at times, straightforward. For example, the multilayer degree is obtained by summing up the vector containing the nodal degrees in each layer. Multilayer and multiplex versions of local and global metrics have also been implemented: clustering, eigenvector centrality, eigenvector and others. The interpretation of multilayer network metrics is often the same as for their 'classic' counterpart, with the obvious complication derived from their more complicated structure (an in-depth discussion on the mathematics involved and the metrics developed for multiplex networks can be found in Cozzo *et al.* [2015], De Domenico *et al.* [2013, 2015] and Nicosia & Latora [2015]).

Worth mentioning is the importance of two specific metrics that can define nodal and overall multilayer network characteristics: participation coefficient and interlayer assortativity. The participation coefficient describes how many connections a node has within a layer with respect to those it has in all other layers. Given the nature of a socioeconomic system (e.g. a tourism destination), for example, a specific organisation could be central to the flow of information to customers (tourists) but may not be important when it comes to encouraging transactions between different businesses. Another entity may not be very relevant for each type of relationship (i.e. in each layer), but may have a more prominent role when all relationships are taken into account. Interlayer assortativity allows us to analyse whether a node highly connected in one layer is also highly connected in another layer (Baggio *et al.*, 2016). The use of multilayer networks to analyse complex systems has already shed light on some interesting properties that cannot be inferred by looking at an aggregation of the network alone in economics, ecology, neuroscience and social-ecological systems research.

As often happens, different methods can be combined to improve our ability to find an answer to a research question or a solution to a problem. A multilevel network classifies nodes into different layers or levels. The edges thus represent links between nodes in each specific layer and between each layer and we can extend the notion of motif to the micro-configurations formed across the different layers.

Figure 4.4 shows an example. Let us assume that the network is composed of organisations of a certain type (A, B, C, D, G, F) that are linked to some other local-level businesses (1, 2, 3, 4). If we look at the overall network, we can assess the overall relevance of each agent in the system. However, if we want to infer potential issues, for example arising from miscommunication, or potential conflict areas, we can analyse the network by looking at specific motifs and their presence within the network. In this case, we can decide to extract and assess four motifs and interpret them in this context.

According to recent studies, the presence of *motif a* promotes the resilience of the system as it represents organisations collaborating on some business. *motif b* can indicate a danger of conflict, as multiple organisations that are not collaborating are relating to another actor; the same goes for *motif d*, which can be detrimental to the overall resilience of the system, as organisations with connected businesses are not collaborating. Finally, *motif c* represents the importance of organisation G

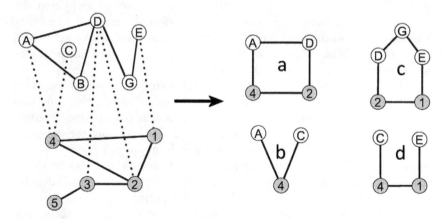

Figure 4.4 Layered network and examples of micro-configurations (motifs)

as a broker between two agents. In other words, how G mediates (collaborates and allows information exchanges) between D and F will have an important impact on the overall ability of the system to perform and be resilient to future changes. Once the configurations of interest have been identified, the application of an ERGM provides the weight of the presence of the chosen motifs and allows an assessment of their overall impact on the system.

Example applications

A naturally multilayered network is the transportation system in a city. Citizens and tourists use different means to move around their city. They may use a combination of buses and metro, availing themselves of the interconnections between the two systems. The two are frequently studied separately, but the different modes of transportation are not independent, and their coupling can be critical to the mobility of people. Moreover, a better understanding of their combined characteristics can be crucial for urban planning decisions, for example in selecting routes to monitor or rearrange during mega events.

A recent study took this combined approach to investigate the properties of a multimodal transportation network (Baggag, 2017). The city used as an example for the model is London. The two layers were the two systems' stations (bus and metro stations) and the edges were the route segments connecting them. An interlayer link was set if there was a walking distance (less than 100 m) between a bus and a metro station.

The movements of people were simulated by using random walks on the multilayer network. At each step, walkers could remain where they arrived, move to another station of the same system or switch to the other transport system if there was a connection. To assess the mobility features, the authors determined the coverage of these walks defined as the expected value of the number of steps needed to reach all the nodes in the network, regardless of the layer.

One of the important objectives in such a study is to assess the overall robustness of the multimodal system. In other words, they imitated situations in which one or more of the tracks was unavailable or unusable (street accidents, traffic jams, etc.) and checked how the coverage changed and the level of disruption for which the overall system became inoperative. This is simply done by iteratively removing edges from the multilayer network and calculating the new coverage, which will depend on the number of links removed. In the case examined, the study found that the London multimodal transportation system has good robustness: the removal of 70% of its edges led to a 20% loss of coverage.

The method is a relatively simple application of the multilayer network model. It is effective in extracting several features of a combined transportation system (provided the routes' data are available) and is easily replicable or extendable to any area of interest.

A second interesting example analysed the multiple relationships that bind different online sources to identify points of interest (POI) at a destination.

The use of web platforms for accessing several services such as itinerary planning, comments, reviews, ratings and booking accommodations or searching documentary material on venues and places is well established in the community of travellers. The amount of information that can be recovered, however, is ever growing and can overload the user, often generating more confusion than help with choices because of the heterogeneity of the sources that limits the ability to extract useful information.

A work by Interdonato and Tagarelli (2017) set out to solve this issue by suggesting a way to build a reliable POI database containing places such as monuments, shops, restaurants, attractions and other locations of interest to a visitor. At the core was a multilayer network based on elements that can be recovered from three well-known online platforms: Google Maps, Foursquare and Wikipedia. The interest for the modeler here was that the network considered was made of layers that contained different 'objects' linked by different types of connections originating from the different features of the data used in the three platforms, where the only common element across the three was the geographical location.

The first layer was derived from Google Maps. Once a destination was selected, starting from an initially chosen POI, typically a popular or well-known attraction (seed), a crawler choses the POIs that are within a certain distance from the first seed and are on the list of those suggested by the application. The process was iteratively repeated for each POI found, until a maximum distance from the origin was reached. These were the nodes of the first layer. A link between two nodes was defined and weighted using the minimum spatial distance and time between them, for different modes of transport: pedestrian paths, public transport and private transport. The second layer was recovered from Foursquare. Starting from the same POI as used for Google Maps, the subsequent points were selected from the platform's function of recommending a set of POIs to visit from the input POI. Here, too, the process was repeated recursively. Links were weighted based on user ratings and the comments on the POIs chosen. The Wikipedia layer contained the pages that described the POIs found in the previous layers (if they existed) and the edges were the existing hyperlinks connecting two POIs, weighted by a measure that expressed the content affinity between the wikipages. The interlayer links were set between the corresponding POIs existing in the different layers.

The analysis was basically a modularity analysis in which a specific algorithm, valid for multilayer networks, was used. The communities found in this way had the obvious interpretation of an accumulation of interest by visitors.

Subsequently, with many characteristics collected about the different POIs, the composition of these clusters was inspected, including important information about the interests of the tourists visiting the destination. Applying the proposed method to some Italian cities, for example, they found that the largest interest was in monuments and 'food and drink' locations, not surprising given the locations chosen. The cities, however, showed different compositions of interest. For example, in Florence and Milan, several communities were characterised by the predominance of shops, while in other cities, such as Rome, monuments and restaurants were the most frequent types in the clusters uncovered.

The method proposed is a good example of how a combination of different techniques, on a common base such as a multilayer network analysis, can provide quite interesting and valuable outcomes.

Multilayer networks do not need to have 'consistent' layers or be the basis for consistent processes. A study by Granell et al. (2013), for example, examined the interrelation between two different processes: the spreading of an infection and the diffusion of information to prevent it.

This setting represented a realistic situation in which an epidemic spread across a network of real contacts, and the information about the disease circulated in a set of virtual contacts between the same individuals (e.g. an online social network). This is a multiplex network where two diffusion processes unfold and influence each other. The analysis captured the evolution of the epidemic thresholds and how they were affected by the topological structure of the system. Interestingly, the critical point for the beginning of the epidemic had a critical value defined by the awareness dynamics and the topology of the virtual network. In other words, the diffusion of information on the spread of the disease in the 'real' world had a much greater influence than previously supposed.

The nodes in the layers can represent different entities such as social agents (people, households, etc.) and ecological entities (animals, food sources, ecological systems). Their interactions can then help simulate different scenarios and better understand the interrelations when, for example, disturbances in the natural environment arise and how they can affect the social (and economic) systems, or how the strategic decisions made by the social system (local managers or actors) can influence the availability of ecological resources (food, water, land) (Baggio *et al.*, 2016; Baggio & Hillis, 2018).

Artificial Intelligence Developments: Deep Learning Systems

Deep learning (DL) is a specialised area of machine learning (ML) and refers to algorithms inspired by the structure and function of the brain, called *artificial neural networks* (ANNs). It is thought to be the closest technique to the original goals of the wider field of artificial intelligence, which aims to develop machine systems that mimic the cognitive functions of the human brain.

The architecture of DL (with which the concept of an ANN has gained public attention) has been applied in computer vision, in the automatic recognition of the spoken language, in the processing of natural language and in audio recognition. DL can be defined as a class of ML algorithms that uses various levels of cascading non-linear units to perform feature extractions and transformation tasks. Each successive level uses the output of the previous level as its input.

Essentially, DL is an application of ANNs to some data analysis problem. An ANN is a set of software elements built to resemble a biological neural network. More than a single algorithm, an ANN is a collection of different connected ML processes that work together to handle complex data inputs. Each connection, like synapses in a biological brain, can transmit a signal from one neuron to another. The receiving neuron can process the signal and further signal other connected neurons forming a chain of computational units. They process data labelling or cluster raw input. An ANN is a type of clustering and classification machine. ANNs help group unlabelled data based on similarities discovered in the features of example inputs.

A node of the network is a small algorithm built similarly to a brain's neuron, which fires (transmits a signal) on meeting an appropriate stimulus. In practice, a node combines input from the data with a set of coefficients, or weights, that either amplify or reduce that input, assigning a meaning to the input based on the task the algorithm is learning. In other words, weights are assigned so that the calculated output is the closest possible to the pre-labelled output. These input-weight products are summed, and the sum is passed through an activation function to determine whether and to what extent that signal should progress further through the network to shape the final outcome. If the signal passes through, the neuron has been activated (Figure 4.5).

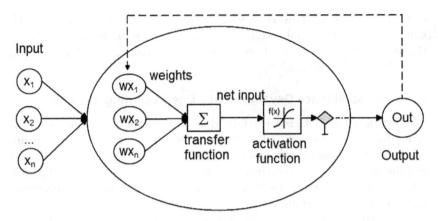

Figure 4.5 Representation of an artificial neuron

In its simplest form, the whole neural network is thus composed of a series of layers: the input elements, the neural layer and the output. The neural layer is usually called the *hidden layer* (Figure 4.6).

The hidden layers can be replicated. They are used for building more levels of abstraction that enhance the overall capabilities of the ANN. The multiple layers are recalculated iteratively using a back-propagation technique for updating the weights of every node. These multiple levels result in much better *learning* for solving the complex problems of

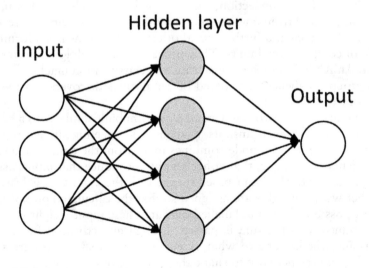

Figure 4.6 A simple neural network

pattern recognition precisely because at each intermediate level they add useful information and improve the reliability of the output. In other words, each layer learns to become specialised and activates when it detects specific features.

Thus, an ANN is a supervised system in which the learning phase consists of the calculation of the weights to be used in the neurons' functions in order to optimally render the pre-labelled output. It is the most advanced reinforcement learning algorithm successfully employed in many areas.

Although an ANN requires large computational capabilities, its performance improves with increasing amounts of data, while usual ML applications, once they have reached a certain level of accuracy, are no longer scalable even if more examples and training data are added. In ML systems, the features of a given object are selected manually to create a model, while in DL systems, the extraction of features takes place automatically: the neural network learns autonomously how to analyse raw data and how to perform a task.

The large computational capabilities needed by DL systems are not usually accessible to individual companies. Most large technology companies, however, offer the possibility to use these systems on a pay-per-use basis (see references in Appendix 2 dedicated to software).

The usual application areas for DL systems are image, voice and text recognition and classification. Additionally, these classification capabilities are employed in building 'artificial' objects from written pieces or paintings or voices, in developing automated translation for several languages and in driving systems for unmanned vehicles.

In recent years, the evolution of these systems has provided an incredible series of results. Image recognition, for example, has reached an error rate of about 2.3% when the human limit is about 5% (Hu *et al.*, 2018). One of its most interesting applications, however, is as a helpful tool for specialised tasks. For example, the combination of advanced ANNs with the competence of a trained pathologist has achieved an incredible 0.5% error rate in the recognition of women's breast cancer (Wang *et al.*, 2016).

In the tourism domain, applications are available that combine, for example, image recognition capabilities with augmented reality in the design of tourist attraction guide systems (Zhou *et al.*, 2019), and software and mobile apps that provide reasonable translation services, their accuracy constantly improving.

DL systems, in particular those for voice recognition, are now commonly employed in robotic devices, such as humanoid robots that are starting to appear in hotels all over the world and that help, typically, concierges and front-desk employees to answer customers' questions and requests.

Example applications

A couple of examples will better show the power and usefulness of a DL approach to the solution of important issues in the tourism domain.

The first example concerns the long-debated problem of forecasting demand. It is a highly investigated area in which a wealth of methods have been designed with mixed results (see e.g. Peng *et al.*, 2014; Song *et al.*, 2010). In a recent case, a DL approach was used not only to improve the reliability and accuracy of the forecast, but also to identify relevant features that would allow an understanding of the relationships between the different tourist factors involved and how they related to the volume of arrivals (Law *et al.*, 2019). The proposed framework availed itself of the online searches performed by tourists and combined these with the more 'traditional' arrival data, thus providing a good model for merging old and new data, which revealed the flow component that is usually under-evaluated because of a lack of appropriate analysis techniques (De Cantis *et al.*, 2015).

The input data included the available monthly measurements of arrivals for the origin markets of interest. This input was then combined with analogous monthly search volumes of items (search keywords) related to the destination. The resulting input was fed into a DL system. The model's training was performed using some built-in mechanism of feature engineering, that is, it extracted the most relevant features from the raw data set using the temporal behaviour of the different items. Thus, the system was able to automatically select a set of influential factors and determine the lag order of time-series sequences.

The resulting trained model represents the temporal relationship between a variety of forecasting factors and the arrival volumes. The weights of the neuron links scores can be applied to determine which original factors have the most influence. Remarkably, the whole process is automatic, and no manual selection or extraction attribute is required. The process uses a walk-forward validation. At each step, the training data 'walk' by one month, and the model is trained to make a forecast for the following month. Then, the actual demand value for the next month from the test set is made available for the next month. A series of traditional forecasting methods is used in parallel to assess the goodness of the DL approach. This comparison shows that the DL architecture performs much better in all accuracy measures.

The second example deals with opinion mining and the sentiment analysis of textual records recovered from the queries and response activities of hotel customers and managers (Martín *et al.*, 2018). As is well recognised today, this is an essential topic as it allows solutions to be crafted that can optimise promotional campaigns, enhance products and services offerings, improve internal operations and, ultimately, increase customer satisfaction, so important in generating value. Moreover, the comments that tourists publish online are increasingly used in the decisions of potential customers, thus proving an essential element in the image of a company or a destination. Reliable and consistent information is needed to understand and foresee client behaviour. This work develops this idea by testing several methods to analyse the reviews that clients provide using digital platforms on the services and products they received.

The sources of the reviews used were Booking.com and TripAdvisor. com platforms. An automatic script gathered the comment texts with their main attributes (date, score, location, etc.). The texts were then cleaned and standardised using text analytic software to prepare them for the DL system. This involved transforming the comments into fixed length vectors of unique words (to allow for different lengths, individual vectors, if needed, were padded with null elements). The scores assigned were used for labelling the reviews, thereby obtaining the training data needed.

Different neural networks models with different filtering techniques were used to find the best possible accuracy in predicting the features of positive and negative reviews. The best performing assembly was able to reach an accuracy of 89.2% (based primarily on their best prediction of positive results).

The results of the implementation used also showed that the best findings were able to classify positive comments. Once trained, this classifier can be further improved by submitting new data and can become the basis for a combination of tools and dashboards providing updated information.

Concluding Remarks

ERGMs and multilayer network models are only two of many new, advanced developments in network science. Many other sophisticated modelling techniques and associated simulation methods are being proposed almost daily in this field, which is probably the most popular approach to the study of complex systems. Its popularity in the tourism domain is increasing due to the many successes obtained in explaining and predicting many phenomena of both theoretical and practical interest.

The same could be said about the recent progresses in artificial intelligence. DL models have proved to be incredibly effective in simplifying many tasks. Their diffusion is supported by a wealth of applications that are proving to be very effective in many domains, of which tourism is one. Even if still in their embryonic stage, researchers, practitioners and software producers are increasingly adopting these techniques. They are paving the way for technological innovation that will have a great impact on the tourism sector. Their potential is very promising and there is little doubt that they will greatly change the way the whole domain evolves.

References

Aleta, A. and Moreno, Y. (2019) Multilayer networks in a nutshell. *Annual Review of Condensed Matter Physics* 10, 45–62.

Baggag, A., Abba, S., Zanouda, T., Borge-Holthoefer, J. and Srivastava, J. (2017) A multiplex approach to urban mobility. In H. Cherifi, S. Gaito, W. Quattrociocchi and A. Sala (eds) *Complex Networks & Their Applications* (pp. 551–563). Cham: Springer.

Baggio, J.A. and Hillis, V. (2018) Managing ecological disturbances: Learning and the structure of social-ecological networks. *Environmental Modelling & Software* 109, 32–40.

Baggio, J.A., BurnSilver, S.B., Arenas, A., Magdanz, J.S., Kofinas, G.P. and De Domenico, M. (2016) Multiplex social ecological network analysis reveals how social changes affect community robustness more than resource depletion. *Proceedings of the National Academy of Sciences* 113 (48), 13708–13713.

Cozzo, E., Kivelä, M., de Domenico, M., Solé, A., Arenas, A., Gómez, S., Porter, M.A. and Moreno, Y. (2015) Structure of triadic relations in multiplex networks. *New Journal of Physics* 17, art. 073029.

De Cantis, S., Parroco, A.M., Ferrante, M. and Vaccina, F. (2015) Unobserved tourism. *Annals of Tourism Research* 50, 1–18.

De Domenico, M., Solé-Ribalta, A., Cozzo, E., Kivelä, M., Moreno, Y., Porter, M.A., Gómez, S. and Arenas, A. (2013) Mathematical formulation of multilayer networks. *Physical Review X* 3 (4), 041022.

De Domenico, M., Sole-Ribalta, A., Omodei, E., Gómez, S. and Arenas, A. (2015) Ranking in interconnected multilayer networks reveals versatile nodes. *Nature Communications* 6, art. 6868.

Granell, C., Gómez, S. and Arenas, A. (2013) Dynamical interplay between awareness and epidemic spreading in multiplex networks. *Physical Review Letters* 111 (12), art. 128701.

Hu, J., Shen, L. and Sun, G. (2018) Squeeze-and-excitation networks. In *018 IEEE/CVF Conference on Computer Vision and Pattern Recognition Workshops (CVPRW 2018). Proceedings of the IEEE Conference on Computer Vision and Pattern Recognition, Salt Lake City, UT (18–22 June)* (pp. 7132–7141). IEEE.

Interdonato, R. and Tagarelli, A. (2017) Personalized recommendation of points-of-interest based on multilayer local community detection. In G.L. Ciampaglia, A. Mashhadi and T. Yasseri (eds) *Social Informatics* (pp. 552–571). Cham: Springer.

Kivelä, M., Arenas, A., Barthelemy, M., Gleeson, J.P., Moreno, Y. and Porter, M.A. (2014) Multilayer networks. *Journal of Complex Networks* 2, 203–271.

Law, R., Li, G., Fong, D.K.C. and Han, X. (2019) Tourism demand forecasting: A deep learning approach. *Annals of Tourism Research* 75, 410–423.

Lazega, E. and Snijders, T.A.B. (2016) *Multilevel Network Analysis for the Social Sciences.* London: Springer.

Lozano, S. and Gutiérrez, E. (2018) A complex network analysis of global tourism flows. *International Journal of Tourism Research* 20 (5), 588–604.

Martín, C.A., Torres, J.M., Aguilar, R.M. and Diaz, S. (2018) Using deep learning to predict sentiments: Case study in tourism. *Complexity* 2018, art. 7408431.

Nicosia, V. and Latora, V. (2015) Measuring and modeling correlations in multiplex networks. *Physical Review E* 92, art. 032805.

Peng, B., Song, H. and Crouch, G.I. (2014) A meta-analysis of international tourism demand forecasting and implications for practice. *Tourism Management* 45, 181–193.

Pržulj, N. (2007) Biological network comparison using graphlet degree distribution. *Bioinformatics* 23, e177–e183.

Robins, G.L., Pattison, P.E., Kalish, Y. and Lusher, D. (2007) An introduction to exponential random graph (p*) models for social networks. *Social Networks* 29, 173–191.

Song, H., Li, G., Witt, S.F. and Fei, B. (2010) Tourism demand modelling and forecasting: How should demand be measured? *Tourism Economics* 16 (1), 63–81.

Wang, D., Khosla, A., Gargeya, R., Irshad, H. and Beck, A.H. (2016) Deep learning for identifying metastatic breast cancer. See http://arxiv.org/abs/1606.05718 (accessed November 2018).

Wang, P., Sharpe, K., Robins, G.L. and Pattison, P.E. (2009) Exponential random graph (p*) models for affiliation networks. *Social Networks* 31, 12–25.

Wang, P., Robins, G., Pattison, P. and Lazega, E. (2013) Exponential random graph models for multilevel networks. *Social Networks* 35, 96–115.

Wäsche, H. (2015) Interorganizational cooperation in sport tourism: A social network analysis. *Sport Management Review* 18 (4), 542–554.

Zhou, X., Sun, Z., Xue, C., Lin, Y. and Zhang, J. (2019) Mobile AR tourist attraction guide system design based on image recognition and user behavior. In *Proceedings of the 2nd International Conference on Intelligent Human Systems Integration (IHSI 2019), San Diego, CA (February 7-10)* (pp. 858–863). Heidelberg: Springer.

5 Choosing a Modelling Method

In the previous chapters, we discussed how tourism systems are complex adaptive systems, and as such should be analysed and modelled. We then introduced models and modelling and detailed how to build models, focusing on the different techniques widely used to analyse tourism systems and often, more widely, complex systems, along with some selected, newer and more advanced techniques to analyse, explain and possibly predict the behaviour of tourism systems. Given the wide variety of modelling techniques available, however, important issues remain: how to choose the appropriate modelling technique and what decisions should dictate the modelling technique we should use.

A model is a simplified representation of reality, and such simplifications essentially consist of discarding those details that are not strictly necessary and allowing the focus on aspects we want to understand. Simplifications are based on assumptions that may hold in some situations but may not be valid in others. This implies that a model that explains a certain situation well, may fail in another situation. No matter the technique used, assumptions must be checked before starting the modelling process.

This statement is not only intuitive but it is also supported by some rigorous deductions known as *no free lunch* (NFL) theorems. The underlying motivation is that there is always a bias in our representations of the observations made when studying a phenomenon or a system and thus in the models we build. In his *Treatise of Human Nature*, David Hume (1739: Book I, Part III, Section XII) states: 'even after the observation of the frequent or constant conjunction of objects, we have no reason to draw any inference concerning any object beyond those of which we have had experience'. Restated by Popper (1969), the idea has been formalised in the field of machine learning (ML), but can be well extended to modelling practice in general (Mitchell, 1980; Schaffer, 1994). In their most known form, the NFL theorems were formally demonstrated by David Wolpert (Wolpert, 1996, 2001; Wolpert & Macready, 1997).

In short, an NFL theorem can be summarised as: any assumptions made produce a bias in a model, but without assumptions, any model would have no better performance on a task than if the result was chosen at random. In other words, how well a model will do is determined by how aligned it is with the actual problem at hand, and there is no universal better model. It is always possible to find a set of variables where one model performs better (in speed, computational cost, accuracy, etc.) than another model, but if a model performs better in a certain situation, it will do badly in all others. Averaging over a large number of settings, all models are equivalent. Therefore, depending on the problem, it is important to assess the trade-offs between the speed, accuracy and complexity of different models and algorithms and find a model that works best for that particular problem.

First and foremost, the appropriate modelling technique depends on the type of problem and research question we want to address. In fact, different research questions, different problems and different interests may favour one modelling technique over another. A second important discriminant when choosing the appropriate modelling technique refers to data, both in terms of types and availability, related to the problem we want to address. A third important decision factor is time/cost and available cyber infrastructure – time/cost to assess the problem, time to collect the data needed, time to implement and analyse a specific model. Lastly, the hardware and software tools available, often those known to the modeler, play an important role and may, at times, heavily condition the choices.

These factors – problem statements, data, resources and time – are the key determinants when it comes to deciding the technique to adopt. Through examples, the remainder of the chapter will showcase in detail how these three decision factors determine the appropriate modelling technique to use.

The Complexity of Models and Simulations

It is usually suggested that a simple model is preferable to a complex one, since modelers may face problems on how to state, validate and understand the results when models become too large and complicated. Moreover, despite the fairly good computational power of modern hardware, some models may require long execution times. Complexity, when applied to models, may look like an intuitive concept, but there is no general or accepted definition. Often it is mistaken with 'level of detail', especially when socioeconomic systems are considered.

In this case, we can distinguish between the concepts of complex and complicated (Sun *et al.*, 2016). While complicatedness is directly related to the level of detail used in a model, complexity, in the case of a computational model, can refer to the overall behaviour and the capacity to capture the complex dynamics of the object of study.

The complexity may increase for a number of reasons, some related to human factors and others of a technical nature. As Chwif *et al.* (2000) report, human factors can be traced to a show-off factor, the desire to show how good a modeler is in devising complex systems, pushing to include as much data and parameters as possible when unsure of the model's real needs, and to the modeler's erroneous trust in the computational power available. From a technical point of view, models grow increasingly complicated when their modeler lacks an understanding of the real system being modelled, or is unable to correctly model the problem (wrongly formulated conceptual model) and set clear simulation objectives or, rather obviously, fails to correctly translate a conceptual model into a computerised one.

Simple models have the undoubted advantage of being easier to implement, validate and analyse. They are also easier to modify when the conditions, hypotheses or knowledge of the problem analysed change, and the simulation time is shorter, thus allowing more combinations to be explored and to quickly obtain results, even if approximate.

However, high simplicity can also have its disadvantages. There may be the problem of validity when an oversimplification leads to ignoring some important elements or parameters, or an issue with the reduction in scope when considering only some seemingly relevant aspects of the object of study that for its complexity cannot be easily reduced (see Chapter 1). A too simple realisation can also be difficult to understand, since the abstraction level used by the modeler can be poorly intelligible to others, raising doubts about the outcomes and affecting the credibility of the implementation.

In essence, the relationship between a complication in a model and its perceived validity or confidence in its outcomes can be drawn as in Figure 5.1, where the relationship between confidence and complication is intuitively sketched. In addition to perceptions, however, the issue of finding the optimal balance in the complication of a model is a significant issue.

As already remarked, modelling and simulation is more an art than a science, and although possessing a wealth of rigorous methodological

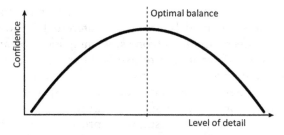

Figure 5.1 Balancing the confidence and complication of a model

paths, a modeler's experience influences the quality and the reliability of the modelling activity. Here, we report a number of suggestions based on the literature and our own experience:

- Start with the simplest possible design and implementation. This will allow feasibility to be checked and, when well-conceived and correctly implemented, can serve as a baseline for further developments.
- Slowly add more details, controlling for effects with respect to the baseline. A good practice is to define at the start some kind of objective function that represents the goodness of the model and use this for comparison, also taking into account the resources (computing power, execution time, etc.) needed for improvement. A balance between the two is crucial for optimal efficiency and effectiveness.
- If possible, adopt a hierarchical modelling technique, in which macro-blocks with their cumulate interactions are included first. Then, increase the level of detail for these blocks trying to balance the different specification levels. If possible, or required for overly complicated models, reduce the scope of the model assigning a priority to the issues under consideration and start from the most relevant issue. This activity may require a deeper analysis of the phenomenon and a more realistic perspective.

Problem Statement

When choosing the appropriate modelling technique, the 'problem statement' is probably the most important element and should be seriously considered. Certain research questions may constrain the choice of modelling technique to implement. If we are interested in feedback and the dynamics of a system, we may need to resort to modelling techniques such as agent-based modelling or system dynamics, rather than statistical models. On the other hand, if our problem statement relates to inferring a causal relationship between variables, we may prefer to implement and develop statistical models. If we are interested in how structural properties may affect specific outcomes, we may look towards network methods and tools. If we intend to assess differences between 'groups' or outcomes, we may again resort to specific statistical models in order to test clear hypotheses. Then, depending on the specific data we want to analyse and on some of their specific features, we might be forced to use ML algorithms, as standard statistical procedures may be unable to provide the answers to the problem we are trying to address.

However, independent of the problem statement we have formulated and are interested in, it is always good practice to start with a conceptual model. Albeit the conceptual model is, as explained in Chapter 3, a modelling technique *per se*, it is often the basis for other model types. As already stated, any model is an abstract representation of a real-world phenomenon or system and, as such, conceptual sketching assumes a central function for the logical consistency of the whole process. Furthermore, due its characteristics, a conceptual model allows focusing on the elements that will enter the model and the simulation, without being sidetracked by some technical aspect or detail that could produce a bias or hide other important elements or variables.

For example, let us look at the different problem statements that underlie modelling decisions based on the examples presented in Chapter 3. As stated, one starts with a conceptual model that represents a way to relate different elements of the system of interest.

The conceptual model, however, also depends on the specific problem statement we want to address. In the statistical model example presented in Chapter 3, we are interested in the abstract, how specific variables/ factors, namely age, country of origin, culture, climate, accommodation quality, etc. (independent variables) affect a variable of interest, namely international tourism satisfaction (dependent variable). Given that we are interested in inferring a relationship between variables, we should rely on statistical models. This is also true when we want to assess how past events are influencing current events (time-series analysis). Numerous books and papers address the issue of what type of correlation, regression, time-series analysis or hypothesis test we should use depending on the nature and structure of the data that are at our disposal (see e.g. Greene, 2003; Smith, 2013; Veal, 2006).

A time series, for example, is one of the most popular modelling techniques for forecasting future values based on previously observed quantities. In the past decades, the research community has developed a wealth of methods with different levels of sophistication. For example,

Song and Li (2008) examined more than 100 published papers and identified about 50 different methods for analysing a time series. The wide availability of these techniques, implemented with a significant number of software programs, makes it possible to apply different methods to the same set of data, and assess the best possible model for the specific situation under study. This appraisal is usually made by splitting the series into two parts: an initialisation (training) set and a test set. Then, using the model built on the training set, a forecast is made and the results compared to the test set (see e.g. Baggio & Klobas, 2017). Many objective functions can be used for the evaluation, such as the average of the absolute differences between the predicted and observed values (mean absolute percentage error: MAPE). These measures can also be used to optimise a specific method or to better tune the parameters employed. It happens that even the most sophisticated methods may fail to provide reliable results (Tideswell *et al.*, 2001): similar methods may provide different results depending on the specific conditions or environments (Papatheodorou & Song, 2005; Wackera & Sprague, 1998). Moreover, as shown by Smeral (2007) who compared different methods applied to the same series, in many situations there is no advantage to using more sophisticated and complicated models, although under other conditions the use of too simple techniques might hinder important effects. A comparison of different modelling methods is important since it can lead to choosing the most suitable method for the particular circumstances at play and provide the most effective representation of the problem at hand. Generally, however, a principle of parsimony (often expressed as Occam's razor) is the best guide.

Statistical models often assume that we are able to formalise a relationship between the independent and the dependent variable, and that we are able to express such a relationship via mathematical equations. Frequently, statistical models aim to predict future behaviour based on current and past observations, assuming that the relationship between the independent and dependent variable will not change dramatically in the future (i.e. the environment in which the relationship is found is taken as constant for the prediction to be reliable). This is, as Chapter 1 discussed, a strong hypothesis in a complex adaptive environment, and can only be assumed valid for a short period of time, supposing the system we study has a kind of inertia that pushes it along the 'usual' evolutionary path with unmodified characteristics. It must be noted here that methods exist for estimating the window of predictability for a complex adaptive system (Andersen & Sornette, 2005; DelSole & Tippett, 2009a, 2009b) and these should be employed before attempting a forecasting exercise.

In other cases, statistical models seek to assess differences or similarities and quantify uncertainties. However, when such formalisation is not possible, but we are still interested in inferring specific patterns and relationships that occur between variables, we may prefer to resort to techniques akin to ML. This may happen when we are interested in similar problem statements as those apt to be investigated via statistical models, but we are unsure or are unable to formalise such relationships, or when the types (and quantities) of data used are not suitable for more traditional statistical methods. As seen in Chapter 3, for example, ML algorithms are more appropriate when we want to find associations between variables that are not easily quantifiable. However, except in a few cases, both statistical models and ML algorithms are exploited to test specific hypotheses, to classify data, to find associations and similarities or dissimilarities or to quantify uncertainties.

Nevertheless, a researcher may not be interested in inferring a causal relationship, finding correlations and/or assessing uncertainties between variables, but is interested in understanding how entities within a system (and in particular a tourism destination) interact. To follow the discussion presented in Chapter 3, for example, if we are interested in how different accommodations as well as restaurants and bars interact and relate to each other (i.e. exchange information and knowledge, have business agreements, etc.), a researcher may choose to resort to analysing the system via network analytic tools. Network analysis (NA) is a powerful way to characterise the structural properties of a destination, as well to assess which entities in a specific destination may control or significantly influence the flow of information, opinions, finances and so on. In other words, when a researcher is interested in mapping and assessing patterns of relationships between entities, an NA should be the preferred modelling technique. Depending on the research question, more advanced techniques (see Chapter 4) related to NA may be employed, especially when one is interested in assessing more complicated relationship patterns. When a researcher wants to address or attempt to understand the processes that may have led to specific network configurations, or when an assessment of the significance of certain structures or quantities is crucial, the researcher may employ techniques within the exponential random graph model (ERGM) family. Additionally, when the main objective is to investigate the patterns of multiple types of relationships or the effects of interdependent networks, multiplex or multilayer NA methods are preferred.

All these modelling techniques (statistical models, ML and NA) are more or less appropriate depending on the problem statement at hand, but they typically work on 'static' snapshots of the phenomenon or system under study. If the problem statement is concerned with the dynamics – the dynamical characteristics of the system or its evolution – or the problem statement is related to some process unfolding on a specific network, and thus an assessment of the feedback existing between the different components is required, one may resort to different modelling techniques: system dynamics or agent-based modelling, or a mix of different modelling techniques (i.e. modelling processes via system dynamics and agent-based modelling), but integrating such processes with an underlying structure (network) to assess both the individual entity characteristics and the structural properties of a system. Whether it is preferable to employ a system dynamic model or an agent-based model (ABM) depends on whether the problem statement of interest is concerned with general feedback between aggregated entities (i.e. potential adopters and adopters, or the interaction between accommodations and services) or if the problem statement is concerned with the individual entities existing in the system and their interaction (i.e. each single potential adopter, interactions between each single accommodation and service entity). In the former case, one may prefer to employ a system dynamic model, while in the latter case, one may prefer to develop an ABM. From Chapter 3, it is worth remembering that a system dynamic model considers only aggregated major components and defines a connection between them in terms of feedback and feedback loops representing flows that have a causal meaning (i.e. more adopters generate more adopters that, in turn, also generates more adoption), and carving a formal relationship between such aggregated entities is usually needed. On the other hand, ABMs make use of methods that allow modelling the system 'from the bottom up', that is, specifying the individual entities and their interactions in order to assess how specific outcomes emerge from such interactions.

Last but not least, it is important to understand that some problem statements may require multiple modelling techniques, as for example, when a researcher is interested in both spatial interactions between destinations and specific processes that happen within each destination, and in how such processes propagate along the overall system. When more complex problem statements such as the one stated are made, it is necessary to integrate multiple methods. Such integration is often required to better assess tourism systems and, more generally, complex systems. One area where such integration has been successfully implemented is the combination of network models with system dynamics and/or agent-based modelling (see Baggio & Hillis [2018] for an example of a network ABM integration). This is an important point since many of the dynamic processes that have been studied for a long time have almost disregarded the structural features of the environment in which the process unfolds. Disregarding such structural properties may lead to non-valid conclusions as it implies that all entities have a probability to interact with all other entities. It has often been assumed, for example, that 'homogeneous mixing' of a population was affected by some spreading process (infections, viruses, opinions, marketing messages, etc.). In other words, it has been assumed that the population has a random distribution of individual features or connections (Hethcote, 2000). Most recent studies, however, have shown that linkage configurations play a crucial role in the speed and the extent of the diffusion. For example, while it was traditionally assumed that a diffusion process could only start after a critical density of agents had been reached, we now know that this is not valid when the distribution of connections is a long-tailed distribution (exponential or power law) because in these cases a critical condition might be absent and the spread could start and progress in any case (Chakrabarti *et al.*, 2008; López-Pintado, 2008; Pastor-Satorras *et al.*, 2015).

Table 5.1 summarises the relationship between problem statements and the preferred modelling techniques to address them. To summarise, statistical models are apt to assess causal inferences and associations between variables when it is possible to assign numerical values to the variables of interests and where one can formalise such relationships. ML techniques are more suitable for assessing relationships between variables when the number of variables and the amount of data are very large or when the data cannot be numerically expressed (i.e. text) or easily formalised. Network analytic methods are used to assess issues that focus on the structure of pairwise relationships between entities. The system dynamics modelling technique can be employed when the dynamics of a system are of interest and the entities can be aggregated. Agent-based modelling is appropriate to assess the dynamics of a system, allowing for the complete disaggregation of entities and thus centring on modelling individual interactions and gauging emergent patterns. Finally, a combination of modelling techniques is advisable for more complex problems such as evaluating the interplay between the structure and entity characteristics within a system.

The recent work by Wu *et al.* (2019) is a good example of how a combination of modelling methods can help answer a seemingly simple question that has numerous intricate aspects. Wu *et al.* (2019) designed and implemented an ABM for modelling inbound tourism to cities in China. The issue here is the evaluation of possible changes in the flow of tourists who visit Chinese cities, attempting to understand what happens

Table 5.1 Generalised problem statement and preferred modelling technique

Problem statement contains	Suggested modelling technique
Assessing differences between groups (hypothesis testing)	Statistical modelling
	Machine learning
Inferring causal relationship	Statistical modelling
	Machine learning
Assessing relationships between variables	Statistical modelling
	Machine learning
Classifying groups	Statistical modelling
	Machine learning
Text analysis	Machine learning
	Deep learning
Understanding/analysing structural characteristics	Network analysis
Analysing entities most connected or controlling information/financial flows etc.	Network analysis
Analysing micro-configurations leading to a specific observed network	ERGM
Analysing macro-patterns of multiple types of relationships and or interdependent networks	Multiplex/multilayer network analysis
Independency and dynamics at high levels of aggregation	System dynamics
Independency and dynamics taking into account entity heterogeneity	Agent-based modelling
How structure and dynamics interact at high levels of aggregation	Network analysis and system dynamics
How structure and dynamics interact taking entity heterogeneity into account	Network analysis and agent-based modelling

when external perturbations arise, such as the reduced accessibility to a city or the opening of a new access point (e.g. a port) or a modification to the attractiveness of a place. The situation needs to take into account the complicated collaborative and competitive relationships between destinations.

This is a typical complicated issue that requires a multi-modelling approach. The first step is to assess the relevant features of the cities considered. This is done by employing statistical modelling in which the main cities are identified based on their 'turisticity' and the possible preferences of tourists arriving or departing from them, and their itineraries are estimated from the results of a large-scale survey conducted in the traditional way. A network is then built by using the cities as nodes and their transport connections (direct flights or rail links) and embedded in the geographical space. The links are weighted using the number of travellers who travel from one city to another. The attractiveness of the different cities is the outcome of a Delphi study, a semi-qualitative method based on experts' opinions.

The network is then used as a substrate for an ABM. The agents (tourists) move from one city to another with a certain probability. This is derived from a gravity model whose parameters are defined using the data collected on the tourists' arrivals and the results of the survey. Once the model is built, a series of simulations can be used for preparing and analysing scenarios that arise from some modification of the network structure (i.e. opening or closing ports, closing cities or modifying attractiveness) or simulating changes in key agent characteristics. As the authors state, this is a very simplified representation, but even so it is able to convey a variety of interesting outcomes. Once these have been further verified, discussed or commented on, it will be possible to 'complicate' the model by inserting new parameters, changing the way existing ones are determined, for instance regarding the attractiveness, that can also be dynamically determined instead of being statically assigned, and so on.

Data

So far, we have seen how specific problem statements may lend themselves to be modelled differently depending on their focus and on what a researcher wants to achieve or is interested in. However, problem statements are not the only key factor that a researcher needs to take into account when deciding the most appropriate modelling technique to use. Data availability and typology are as important as specific research questions when it comes to choosing a modelling environment.

Obviously, the initial question determines the type of data we need either to collect primarily or to gather from secondary data sources. However, the nature of these data can determine or constrain the type of modelling technique we are able to employ. Here, a clear distinction needs to be made between the type of data available and whether the data are available at all. If the data are not available, then we may need to resort to 'synthetic' data creation via simulations based on known ideal distributions of some characteristics.

For example, the structure of a real social network in a population is extremely difficult, if not impossible, to obtain. However, starting from census and demographic data, combined with other public information and reliable surveys, it is possible to build a synthetic population

(Fumanelli *et al.*, 2012). In a case like this, although direct validation is almost impossible, indirect assessment can be achieved. If we run a simulation of a known process such as the spread of an infectious disease for which we have the relevant data, it is possible to tune the network characteristics so that they generate the most similar behaviour to the 'real' epidemic mechanism.

This synthetic population can then be used for several purposes, for example to study how the image of a destination is formed and communicated, or the best combination of starting points for the spread of a marketing message.

The quantity of records also has relevant consequences. When the amount of data recovered is limited, a researcher will be limited in the choice of modelling techniques. As seen, statistical models, ML (including deep learning) and NA (including ERGM and multiplex or multilayer networks) have certain requirements for or constraints on the quantity of data/items they use. On the other hand, system dynamic models and ABMs can be designed and implemented with fewer data restrictions even in the absence of observed or real-world data, provided a strong theoretical foundation or good expert knowledge exists (see Chapter 2).

If data are available, a researcher should always assess the type of objects (i.e. variables, parameters) to be used, their quality and the reliability of the collection methods (see also Baggio & Klobas, 2017). As stated previously, the research question indicates the type of data required. For example, if we want to model how 'reviews and comments' assess visitations to a specific tourism destination, we will mainly require textual data, and hence the modelling techniques available to us to make sense and assess the patterns we are interested in are limited to ML (and, in some cases, deep learning). By the same token, if our problem statement relates to assessing how structural properties (or a pairwise relationship between entities) influence specific outcomes, we will collect data required for an NA. Thus, while data are a key component of the decision-making process on whether to employ one modelling technique over another, they are also often inextricably linked with the problem statement we want to address. It must be noted here that the availability of certain types of data may also induce some modifications to the original question or impose the formulation of supplementary hypotheses. For example, textual data, with the sentiment they express, may be considered the best possible source for assessing the influence of reviews on some interesting outcomes. If, instead, only images (e.g. photographs taken by tourists) are more easily collectable, following desk verification of the literature, we may introduce the supplementary hypothesis that the amount (or the frequency) of pictures is an indication of interest and appreciation, and use these types of data instead, as they are less expensive to collect and analyse.

Once data are collected, the first key step is to clean and assess them. Much has been written on the problem of evaluating the quality of data collected. Table 5.2 summarises the issue with a synthesis of the main aspects regarding the quality of data (adapted from Baggio & Klobas [2017] to which we refer the interested reader for more details).

Two aspects are important for practical use: the existence of missing values and the presence of outliers. Missing values can affect statistical

Table 5.2 Main features defining the quality of data

Feature	Definition
Accessibility	Extent to which data are available and quickly retrievable
Amount	Extent to which the volume of data is appropriate for the task at hand
Believability	Extent to which data are regarded as true and credible
Completeness	Extent to which data are not missing and of sufficient breadth and depth for the task at hand
Conciseness	Extent to which data are compactly represented
Consistency	Extent to which data are presented in the same format
Ease of manipulation	Extent to which data are easy to manipulate and apply to different tasks
Error-free	Extent to which data are correct and reliable
Interpretability	Extent to which data are in appropriate language, symbols and units and definitions are clear
Objectivity	Extent to which data are unbiased, unprejudiced and impartial
Precision	Extent to which data are measured to the required level of specificity
Relevance	Extent to which data are applicable and helpful for the task at hand
Reputation	Extent to which data are highly regarded in terms of source or their contents
Security	Extent to which access to data is restricted appropriately to maintain their security
Timeliness	Extent to which data are sufficiently up to date for the task at hand
Understandability	Extent to which data are easily comprehended
Value added	Extent to which data are beneficial and provide advantage through their use

Source: Baggio and Klobas (2017).

models and ML techniques differently. Some models can be more sensitive to missing data or may even be inapplicable, although ML techniques are usually more flexible in handling not only large amounts of data but also omitted or misplaced values. Once missing values are treated and the decision has been made on whether they will impinge on the researcher's ability to assess the issue examined, it is necessary to scan the data to identify potential outliers (usually drawing the distribution or using basic descriptive statistics). Outliers, as missing values, can influence the results of the analysis. Once again, however, different statistical models as well as ML techniques can be more or less sensitive to outliers. Finally, in some cases, the distribution of values plays a role. For example, when training an ML model to classify some objects, having multiple classes with an imbalanced data set (i.e. some classes are more numerous than others) makes many conventional algorithms less effective, especially in predicting minority class examples.

The same applies when the distributions of the training set are vastly different from those of the test set. Some techniques exist to solve this issue, but these must be well assessed when interpreting the outcomes (see e.g. Hendrycks & Gimpel [2016] who also make an open source software program available).

Statistical models, on the other hand, are useful for specific problem statements that want to infer the causality of finding specific associations between data. If data are available and they conform to specific structures (often numerical values as well as some variation of a multi-normal distribution are required), we can then employ traditional statistical models. It is worth mentioning that statistical models can be seen as a subset of ML models. The choice of ML model, however, can depend on the input data as well as the required outcome. With respect to the input data, ML techniques requiring supervised learning algorithms (see Chapter 3) are preferable when data can be labelled; however, when data cannot be labelled, one may resort to ML unsupervised learning algorithms and modelling techniques. Finally, when a researcher is interested in optimising a specific fitness function within a specific environment, the preferred ML technique will be a reinforcement learning algorithm (e.g. a deep learning neural network).

The outcome sought may determine the choice of appropriate ML technique. And here, it is clear that statistical models can be thought of as a subset of ML models. If one is interested in the output as a 'quantifiable entity', one employs regression techniques. However, the appropriate regression technique is determined by the data and other assumptions (see Baggio & Klobas, 2017). On the other hand, if one is interested in classifying the output (that is, the output as a 'class'), the ML of choice

Table 5.3 Data type and availability and preferred modelling techniques

Empirical data	Suggested modelling techniques
Not available	System dynamics
	Agent-based modelling
Relational data	Network analysis
Unstructured and non-numerical data	Machine learning
Large amounts of data	Machine learning
Numerical data that can be defined by equations (in relatively small quantities)	Statistical models

will be a classification problem. ML algorithms can also be employed to assess data clustering as well as anomalies in data. Table 5.3 summarises the considerations made so far.

Other Decision Factors

While problem statements and data type and availability have a very important role to play when deciding the modelling technique to employ, other important factors need to be taken into account. Of these, we briefly mention the time and cost of research as well as the processing speed and storage capacity.

If data are already available, often statistical models and ML algorithms as well as NA may offer insights into a specific problem in a shorter time compared to building ABMs and system dynamic models. However, if we take into account the time required to gather and prepare the data to implement ML and statistical and network models, the scales are reversed. For ML supervised algorithms, for example, setting up of a reasonably dimensioned, labelled data set can be a long and laborious task as some thousands of items are needed for reliable application.

Thus, system dynamic models and ABMs may be 'quicker' in providing specific results for the phenomena or the systems we are interested in studying. Time cannot and should not be the primary factor in deciding the modelling technique to adopt, although it can play an important role in defining 'doable' problem statements and thus the adoption of 'doable' modelling techniques.

As already stated, some computerised modelling methods can be quite intensive computationally and require large capacities for storing not only the source data and the results, but also those produced in the intermediate steps. For an image recognition model, for example, and relatively small or simple tasks, a state-of-the-art computer can handle some hundreds or thousands of images per second, but if the complexity of the model increases or the amount of objects to classify increases (as often happens), or when real-time voice recognition is required, then well-dimensioned clusters of machines are required to obtain results in a decent period of time. This is not usually available in 'normal' companies that, of necessity, must resort to a cloud service.

Time and cost can be thought of as having the same effect on the modelling decision-making process. Cost is high when data need to be collected, and cost can be higher if we need to increase the storage capacity and processing speed of the cyber infrastructure; therefore, we rely on implementing and running specific models. Depending on storage capacity, one may or may not be able to employ ML algorithms on very large amounts of data (gigabytes and beyond). At the same time, the processing power may limit the speed at which models can be simulated and results can be made available.

In conclusion, choosing the appropriate modelling technique initially depends on what we are trying to address, the problem statement we are interested in and what that problem statement entails, or the system we want to examine. Second, it also depends on the data available (or collected), and finally, it can depend on other factors such as time, cost, storage and processing speed. Figure 5.2 summarises the connections between the problem statement, data availability and modelling techniques. It assumes that we have the necessary time, money, storage capacity and processing speed to freely choose between different methods.

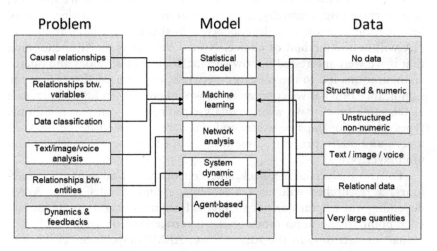

Figure 5.2 Problem statements, data and modelling techniques

To recap the examples portrayed in Chapter 3, if we want to know what determines tourist satisfaction, we are interested in a causal inference between satisfaction factors (such as food, hospitality, shopping, security, etc.) that represent our independent variables, and overall tourist satisfaction (our dependent variable). Further, in this case, all variables can be measured numerically, and we can formalise their relationship by expressing them through mathematical equations. When numeric and structured data are available, the preferred technique would be a statistical model (in Chapter 3, we employed a simple regression model). Also following Chapter 3, if we were interested in the relationship between the other destination visited and the one we are focused on now, the data could be in text format that could be translated into nominal values. However, these data are unstructured, and it is not clear whether tourists at the destination are likely to have previously visited other specific destinations (if any) or which ones. In this case (see also machine learning in Chapter 3), we are interested in a potential causal relationship between an unspecified previous destination and the destination of interest. Since our data are unstructured and are not amenable to be easily 'translated' into a mathematical equation, we resort to ML techniques to assess the association between previously visited destinations and the destination of interest and guess the probability that a visitor to a specific destination will visit the one of interest.

To continue with examples from Chapter 3, we now focus our attention on understanding how the network of relationships in a specific destination reveals the specific properties of the destination we are studying, such as clustering, hierarchy, asymmetry in the ability to divulge, control information or economic dependencies. In this case, it is clear that an NA will be the method of choice. Nonetheless, it is worth mentioning that if we were interested in how specific attributes of nodes and micro-configurations led to the observed network, or we needed to gauge the significance of what we computed, we would employ an ERGM method; at the same time, if we were to analyse the network and keep multiple types of relationships separate, and we were interested in the network macro and micro properties, we would have to resort to multiplex and multilayer NA.

For the analysis of feedback, interdependencies and dynamic behaviours, a researcher will need a system dynamic model or an ABM, depending on whether or not theoretical arguments and expert knowledge point to the importance of the heterogeneity of agents and individual interactions. In the former case, ABMs may be preferred, and system dynamics in the latter case. In both cases, data can be used to calibrate the model (see Chapter 2 on computational models, and Chapter 3). Developing and building ABMs and system dynamic models does not necessarily require empirical observations. However, as detailed in Chapters 2 and 3, the researcher needs to ensure that all the model assumptions, the relationships between variables and the feedbacks and dynamics implemented are reasonable, based on expert or local knowledge, on the literature or, indirectly, on some observed phenomena that can relate to the model in preparation. The adoption of innovations is a clear example of when system dynamic models and ABMs can help shed light on a specific interplay between the dynamics of a system and its topology. If the research question concerns the feedback between adopters, potential adopters and word of mouth, we may resort to model the dynamics of adoption with a system dynamic model. On the other hand, if we have reason to believe that each individual entity in a tourist destination may have different adoption preferences, for example based on different expectations, perceived utility or cost, and value word of mouth differently depending on which other entity they receive the message from, then we may want to design an ABM. ABMs are often used to explicitly consider high heterogeneity in the agent characteristics and high diversity in the interactions between the agents and environments, as well as other features such as structure, dynamics, feedback and adaptation. Finally, if we have data on the evolution of a specific innovation's adoption, we can calibrate both types of models to assess which parameter leads to the observed reality. By the same token, if we do not observe the actual evolution of adoption but have good estimates on costs and expected returns as well as 'advice' value, we can parametrise the model accordingly and assess the results.

Concluding Remarks

Choosing the correct modelling technique is not an easy task; however, the recommendations in this chapter should guide the researcher in making informed decisions. Obviously, once the preferred modelling technique has been chosen, it is important to understand and carefully craft the model. Having a clear problem statement, clear data requirements and availability, as well as clear assumptions on the interaction between the variables and elements of the model is fundamental for assessing real-world phenomena via modelling. However, this may not be enough. Frequently, a 'correctly' designed and implemented model does not imply a 'good' model. Even when we think we have a good model, this may or may not be the case. In all cases, it is important to be as transparent as possible and take the steps to a specific, clear and explicit solution. A description of the model should always be complemented with the code and the algorithms, the choice and the calibration of the different parameters, as well as the data (when possible) used to perform the analysis and on which a model is based.

Finally, as already stated, good experience in these activities is a fundamental requirement in arranging efficient and effective models and simulations.

References

Andersen, J.V. and Sornette, D. (2005) A mechanism for pockets of predictability in complex adaptive systems. *Europhysics Letters* 70 (5), 697–703.

Baggio, R. and Klobas, J. (2017) *Quantitative Methods in Tourism: A Handbook* (2nd edn). Bristol: Channel View Publications.

Baggio, J.A. and Hillis, V. (2018) Managing ecological disturbances: Learning and the structure of social-ecological networks. *Environmental Modelling & Software* 109, 32–40.

Chakrabarti, D., Wang, Y., Wang, C., Leskovec, J. and Faloutsos, C. (2008) Epidemic thresholds in real networks. *ACM Transactions on Information and System Security (TISSEC)* 10 (4), art. 1.

Chwif, L., Barretto, M.R.P. and Paul, R.J. (2000) On simulation model complexity. In *Proceedings of the 32nd Winter Simulation Conference*, Orlando, FL (10–13 December), 449–455.

DelSole, T. and Tippett, M.K. (2009a) Average predictability time. Part I: Theory. *Journal of the Atmospheric Sciences* 66 (5), 1172–1187.

DelSole, T. and Tippett, M.K. (2009b) Average predictability time. Part II: Seamless diagnoses of predictability on multiple time scales. *Journal of the Atmospheric Sciences* 66 (5), 1188–1204.

Fumanelli, L., Ajelli, M., Manfredi, P., Vespignani, A. and Merler, S. (2012) Inferring the structure of social contacts from demographic data in the analysis of infectious diseases spread. *PLoS Computational Biology* 8 (9), art. e1002673.

Greene, W.H. (2003) *Econometric Analysis*. Harlow: Pearson Education Ltd.

Hendrycks, D. and Gimpel, K. (2016) A baseline for detecting misclassified and out-of-distribution examples in neural networks (arXiv preprint:1610.02136). See https://arxiv.org/abs/1610.02136 (accessed May 2019).

Hethcote, H.W. (2000) The mathematics of infectious diseases. *SIAM Review* 42 (4), 599–653.

Hume, D. (1739) *A Treatise of Human Nature*. London: John Noon.

López-Pintado, D. (2008) Diffusion in complex social networks. *Games and Economic Behaviour* 62 (2), 573–590.

Mitchell, T.M. (1980) The need for biases in learning generalizations. In J.W. Shavlik and T.G. Dietterich (eds) *Readings in Machine Learning* (pp. 184–191). San Mateo, CA: Morgan Kaufmann.

Papatheodorou, A. and Song, H. (2005) International tourism forecasts: Time-series analysis of world and regional data. *Tourism Economics* 11 (1), 11–23.

Pastor-Satorras, R., Castellano, C., Van Mieghem, P. and Vespignani, A. (2015) Epidemic processes in complex networks. *Reviews of Modern Physics* 87 (3), 925–979.

Popper, K. (1969) *Conjectures and Refutations*. London: Routledge and Kegan Paul.

Schaffer, C. (1994) A conservation law for generalization performance. In *Proceedings of the Eleventh International Conference on Machine Learning* (pp. 259–265). New Brunswick, NJ: Morgan Kaufman.

Smeral, E. (2007) World tourism forecasting: Keep it quick, simple and dirty. *Tourism Economics* 13 (2), 309–317.

Smith, S.L.J. (2013) *Tourism Analysis: A Handbook* (2nd edn). London: Routledge.

Song, H. and Li, G. (2008) Tourism demand modelling and forecasting: A review of recent research. *Tourism Management* 29, 203–220.

Sun, Z., Lorscheid, I., Millington, J.D., Lauf, S., Magliocca, N.R., Groeneveld, J., Balbi, S., Nolzen, H., Müller, B., Schulze, J. and Buchmann, C.M. (2016) Simple or complicated agent-based models? A complicated issue. *Environmental Modelling & Software* 86, 56–67.

Tideswell, C., Mules, T. and Faulkner, B. (2001) An integrative approach to tourism forecasting: A glance in the rearview mirror. *Journal of Travel Research* 40, 162–171.

Veal, A.J. (2006) *Research Methods for Leisure and Tourism: A Practical Guide* (3rd edn). Harlow: Financial Times/Prentice Hall/Pearson Education.

Wackera, J.G. and Sprague, L.G. (1998) Forecasting accuracy: Comparing the relative effectiveness of practices between seven developed countries. *Journal of Operations Management* 16 (2–3), 271–290.

Wolpert, D.H. (1996) The lack of a priori distinctions between learning algorithms. *Neural Computation* 8 (7), 1341–1390.

Wolpert, D.H. (2001) The supervised learning no-free-lunch theorems. In *Proceedings of the 6th Online World Conference on Soft Computing in Industrial Applications* (10–24 September). See http://www.no-free-lunch.org/Wolp01a.pdf (accessed January 2019).

Wolpert, D.H. and Macready, W.G. (1997) No free lunch theorems for optimization. *IEEE Transactions on Evolutionary Computation* 1 (1), 67–82.

Wu, J., Wang, X. and Pan, B. (2019) Agent-based simulations of China inbound tourism network (arXiv preprint:1901.00080). See https://arxiv.org/abs/1901.00080 (accessed May 2019).

6 Tourism and Hospitality Case Studies

In this chapter, we provide further case studies in the tourism and hospitality domain in which the techniques and methods described in the previous chapters are applied to real situations. The four cases we present focus on (a) assessing and predicting European Union (EU) international tourist flows via an agent-based model (ABM); (b) predicting hotel booking cancellations via machine learning (ML); (c) analysing networks for the importance of networking in hospitality structures; and (d) relating tourism and environmental sustainability via a system dynamic model and an ABM. For each case, we briefly describe the problem statement and how it suggests the specific methods to employ. We then show how such methods are implemented and used, the requirements and the collection of data, the application of the method chosen and the outcomes that derive from it. This chapter is an *applied* guide on how to choose and employ advanced modelling techniques via specific examples both *ad hoc* and taken from the literature.

International Tourism Flows between European Countries

Problem statement

Europe is the most visited region of the world (UNWTO, 2016; WTTC, 2016); tourists within the EU frequently travel abroad and have the advantage of a common currency (nine EU countries do not use the Euro). In this context, it is interesting to assess the basis for the decision-making process of families and individuals when choosing an international destination. In fact, modelling the factors that influence an individual's or family's decision to travel internationally is important for devising policies that have a positive effect on tourist flows. Hence, our problem statement concerns an assessment of the most influential factors in the decision-making process of individuals and families (henceforth called agents) when it comes to choosing international destinations. At the same time, we are interested in how these micro-level decisions affect macro-level patterns of tourist flows. Further, we focus our interest on Europe.

Methods: Choice and use

Assessing individual decision-making is a difficult task. Although excellent experiment and survey techniques exist that can elicit an individual's decisions, such methods are costly and time-consuming. Further, while such methods exist, one would analyse the data from these choice experiments, surveys and interviews via more standard traditional techniques. However, such analysis cannot take into account changes in preferences over time, the effect of individual experiences and how these may modify destination preferences in the future. Thus, the problem statement requires a different modelling approach. More specifically, since we are interested in dynamics (individual decision-making over time), we want to consider agents' heterogeneity as well. Therefore, we decide to implement an ABM (see Chapters 3 and 5). The choice of using an ABM is also favoured by the fact that specific individual data may be lacking or difficult to gather due to cost and time.

Moreover, in this case, we are also interested in assessing individual factors to suggest potential policy changes that may affect a country's overall attractiveness. Hence, while our ABM can be based on a theoretical understanding of decision-making processes, it should also take into account the available data. In other words, to more rigorously assess potential policy changes, the ABM should be empirically calibrated (see Chapter 2 on calibration, fitness and sensitivity analysis, and Chapter 3). In summary, we are building a simple, empirically based ABM that will consider some known key variables in tourist decision-making. The model will evaluate tourist flows for the years 2000 and 2010 and compare the results with the data made available by the UN World Tourism Organisation (UNWTO, 2014).

Data collection and preparation

Given that our model is empirically based, we reviewed the literature and found specific factors that are considered key in affecting agents' decision-making process within the context of international travel. Further, to better mimic the population of interest, we assessed the number of trips each agent would make during a year as well as the number of international travellers and the population of the different countries.

Within the factors affecting decision-making, each agent was able to have full information and assess climatic, cultural, cost and competitiveness differences between their place of origin and the destination of choice. Cost was considered a critical variable for most of our individual decisions (or consumer decisions). The choice of the variables to include was climate, culture, cost and competitiveness (Bigano et al., 2006; Cohen, 2009; Scott et al., 2016; Williams & Baláž, 2015). Finally, prior experience needed to be taken into account as it reduces uncertainty about a destination and, if positive, increases the probability of returning to the destination, while if negative it significantly reduces that probability (Gursoy & McCleary, 2004; Lehto et al., 2004; Oppermann, 2000). The selection of variables and parameters to include in an ABM must stem from the literature or empirical observation (see also Chapters 2 and 3).

Data for the number of trips and international travellers were taken from Eurostat (2016). The model comprised 18,386 agents (each representing approximately 2,000 travellers) distributed according to the number of international travellers for each of the 29 countries examined. Each agent made an average of nine trips abroad in both 2000 and 2010. Data on climate were taken from Amelung and Moreno (2009) who reported on the average monthly values and considered several climate-related elements affecting the quality of the tourism experience. We gathered data related to cultural differences from Hofstede (2001). These data report cultural similarities and differences between European countries. Destination attractiveness was proxied from the tourism competitiveness index and gathered from the World Economic Forum (WEF, 2015). Costs were provided by the evaluations published by Numbeo (2016), a crowdsourced database of the main living costs in different countries (rent, groceries, restaurant prices, etc.). Finally, as experience data were not available and thus would not be driven by the literature, they were used as a parameter that influenced agent choices either by increasing or decreasing the probability of an agent returning to a previously visited destination. The advantage of computational models (and models that do not require data such as ML and statistical models) is that in the absence of data, one can 'vary' specific parameters to assess their importance and, at the same time, the sensitivity of the outcome to the parameters of interest.

Worked example

Once all the data had been gathered, we designed a basic equation that mimicked the decision-making process of each agent in the model.

One important assumption we made at the beginning was that all European countries were part of an initial choice set of places to visit and thus were equally likely destinations (Woodside & Lysonski, 1989). First, each agent randomly chose a country (at the beginning of the simulation). If the country chosen was not in line with the agent's personal preference, the agent chose a different country for a maximum of five times, after which the agent decided not to travel abroad. If the country selected was in line with an agent's preference, the agent would actually choose to travel based on a probability given by

$$P(T_j) = \frac{1}{1 + e^{-(ctr_j + w_{exp}*exp)}} \qquad (1)$$

where $P(T_j)$ is the probability that agent i will travel to country j and

$$ctr_j = w_{clm}*clm + w_{clt}*clt + w_{cmp}*cmp + w_{cst}*cst \qquad (2)$$

where w_{clim}, w_{clt}, w_{cmp}, w_{cst} and w_{exp} represent the weights given by each agent to the following quantities:

- clm = climate difference between country of origin i and country j;
- clt = cultural difference between country of origin i and country j;
- cmp = tourism competitiveness difference between country of origin i and country j;
- cst = cost difference between country of origin i and country j;
- exp = experience of traveling abroad (exp = 0 at t = 0):

$$exp = \frac{1}{1 + e^{-(ctr_j)}} \pm rnd \qquad (3)$$

where rnd = random number drawn from a random uniform distribution in the interval $[0, 1 - (1/1 + e^{-(ctr_j)})]$.

Given that preferences are based on differences within the key variables, we calculated all differences as cosine differences. For clm, cmp and cst, a sign was added, given by the signed difference of the arithmetic means of the countries' values for each origin–destination pair. The sign gives the possibility of choosing a destination with greater or lower levels of the variable of interest. Finally, if exp is greater than a positive experience threshold, agent i will return to country j in the next trip and with probability 0.5 will either retain the experience or diminish the experience by a random quantity between 0 and the current experience level.

Once this part of the model was coded (NetLogo was used, see Figure 6.1), the ABM needed to be calibrated. As this was an empirically based ABM, we needed to assess which parameter combination would be more likely to give the results we observed on the ground. In this way, we started to assess the relative importance of the different factors in determining destination decisions. To do this, we calibrated the model on the observed tourist flows between the different EU countries.

More specifically, calibration of the model was performed on the weights given to the different decision-making components and by the individual agent preference climate and cultural probabilities, as well as the threshold for an agent to be satisfied with a destination and thus return for a subsequent trip. As a starting point, we used the number of travellers abroad in each country (from Eurostat, 2016) and ran the model for nine trips representing the average number of trips made by European international tourists (Eurostat, 2016).

Calibration parameters should always be reported to facilitate the reproduction and transparency of the modelling process. Along these lines, we report that the calibration process used a genetic algorithm: the standard *GreyBinary Chromosome* genetic algorithm from the NetLogo *BehaviourSearch* 1.02, with population size = 50, mutation rate = 0.02, crossover rate = 0.7 and tournament size = 8. We stopped the genetic algorithm after 1000 evaluations of fitness (roughly the *goodness* of the solution). It is important to note here that the fitness was only calculated for new parameter combinations. We rechecked our best fitness 10 times to avoid results driven by chance. We performed 15 different searches.

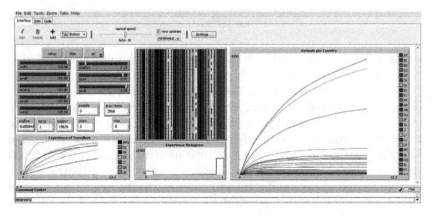

Figure 6.1 NetLogo interface of the model implementation

Fitness is the key variable on which the calibration process was based (see Chapter 2). In the calibration context, recall from Chapter 2 that fitness is basically either the minimisation of 'distance' or the maximisation of 'similarity' between the results of our model given a parameter combination and some benchmark normally given by observed data. In this case, fitness was minimised by the square distance between the observed flow and the simulated tourist flow. This resulted in a fitness function that was the sum of all 29×29 (=841) tourist flow combinations from country i to country j. Formally:

$$f = f_{2000} + f_{2010}$$

where

$$f_{2000} = \sum_{i,j}^{N} (obs_{ij,2000} - sim_{ij,2000})^2$$

$$f_{2010} = \sum_{i,j}^{N} (obs_{ij,2010} - sim_{ij,2010})^2$$

where obs = observed travellers and sim = simulated travellers.

The formula used to calculate the fitness, however, did not allow us to assess how *good* this simple model is. In order to better gauge intuitively the fitness of the model, we normalised the fitness results, f, by the number of potential flows ($N(flows)$ = 841), which allowed simpler recognition of the average discrepancy between the actual and the simulated travel from country to country. Formally:

$$ff = \frac{\sqrt{f}}{N(flows)}$$

Table 6.1 reports the fitness (ff) results, the optimal parameter configuration and the average overall traveller discrepancy for the 10 best fitness metrics recorded.

Overall, the calibration on existing data with respect to the characteristics of the country and to the parameters on the actual data led to good results for the two separate travel flow years (2000 and 2010). However, there was a substantive range in the value of the parameters leading to

Table 6.1 Fitness (*ff*) and average traveller's discrepancy between simulated and observed (Avg ΔPop) data for the top 10 fitness evaluations

Rank	wclm	wclt	wcmp	wcst	wexp	th	pclm	pclt	ff	Avg ΔPop
1	70.09	44.87	62.45	87.31	43.17	0.99	0.12	0.93	14.43	28,865
2	14.68	93.73	59.69	54.5	52.56	0.99	0.16	1	14.97	29,931
3	70.09	61.02	62.45	87.31	43.17	0.99	0.12	0.93	15.49	30,980
4	81.81	3.23	54.68	45.63	28.64	0.99	0.27	0.78	15.96	31,930
5	81.81	0	57.74	45.63	28.64	0.99	0.27	0.78	16.49	32,983
6	81.81	3.23	57.74	45.63	28.64	0.99	0.27	0.78	16.80	33,610
7	20.54	6.31	33.18	39.59	15.58	0.85	0.44	0.97	17.05	34,094
8	3.78	3.23	57.74	45.63	28.64	0.99	0.27	0.68	17.09	34,180
9	23.7	33.31	38.52	55.97	47.57	0.93	0.15	0.83	17.39	34,779
10	98.49	62.31	54.92	0	17.71	0.98	0.12	0.95	17.45	34,908
Average	54.68	31.124	53.911	50.72	33.432	0.969	0.219	0.863	16.31	32,626
Min	3.78	0	33.18	0	15.58	0.85	0.12	0.68	14.43	28,865
Max	98.49	93.73	62.45	87.31	52.56	0.99	0.44	1	17.45	34,908
Median	70.09	19.81	57.74	45.63	28.64	0.99	0.215	0.88	16.65	33,297
StDev	34.83	33.16	9.95	24.76	12.53	0.05	0.10	0.11	1.06	2,111
Range	94.71	93.73	29.27	87.31	36.98	0.14	0.32	0.32	3.02	6,042

Note: *th* = positive experience threshold; *pclt* = probability of individual preferring culturally similar countries; *pclm* = probability of travellers preferring hotter climates.

the best fitness, hinting at the possibility that different factors may be substituted.

To better and more thoroughly assess the fitness of the models depicted in Table 6.1, we report differences in individual travel flows between countries, of observed versus simulated data in 2000 and 2010 for the best ranked model in Table 6.1 (see Figure 6.2). Figure 6.2 showcases potential problems given the complexity and specificity of individuals from different countries. For example, while in the majority of cases the difference between observed and simulated data from country i to country j was very small (near 0), there were some cases in which there was a very high discrepancy between the observed and simulated data. For example, individuals travelling to Finland, Greece and Italy were severely underestimated by the model in both 2000 and 2010. The same can be said of individuals travelling to other destinations from Romania and Austria. Generally speaking, according to Figure 6.2, the underestimation in 2000 led to a further increase in underestimation in 2010. This increase was likely due to experience in international travel. In all the best fitness configurations, experience needs to be extremely positive to warrant a return. That is, loyalty does not seem to play a key role ($th = 0.99$ in 7 out of 10 cases, when max experience =1). Note that two out of the three models ranked in the top 10 parameter configurations have the same values except for culture (see Table 6.2, models ranked 1 and 2). Similar parameter configurations are also present in models ranked 4, 5 and 6.

To assess the robustness of the resulting parameter configuration in explaining the observed tourist flows, we performed a sensitivity analysis

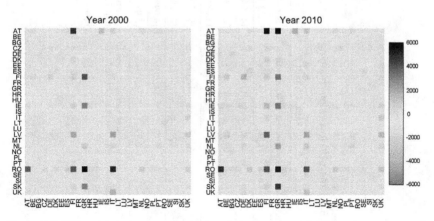

Figure 6.2 Difference between observed and simulated tourist flows from country i to country j

Table 6.2 Fitness (*ff*) loss when changing parameters by 10%

Parameter	Rank = 1, best *ff* = 14.43			Rank = 2, best *ff* = 14.97			Rank = 3, best *ff* = 15.96		
	Par values	Δ*ff*, par +10%	Δ*ff*, par −10%	Par values	Δ*ff*, par +10%	Δ*ff*, par −10%	Par values	Δ*ff*, par +10%	Δ*ff*, par −10%
wclm	70.09	15.32%	15.23%	14.68	6.81%	9.91%	70.09	9.65%	10.31%
wclt	44.87	11.04%	14.29%	93.73	4.12%	17.11%	61.02	9.18%	8.52%
wcmp	62.45	73.49%	22.53%	59.69	−0.12%	28.11%	62.45	55.39%	24.57%
wcst	87.31	14.25%	15.08%	54.5	20.04%	6.02%	87.31	9.10%	10.81%
wexp	43.17	11.47%	15.33%	52.56	11.59%	5.26%	43.17	6.63%	11.55%
th	0.99	72.81%	170.77%	0.99	64.20%	149.73%	0.99	60.64%	152.48%
pclm	0.12	14.23%	16.09%	0.16	6.87%	9.01%	0.12	8.95%	10.57%
pclt	0.93	23.61%	6.91%	1	19.86%	10.81%	0.93	16.69%	2.68%

Note: Par values = parameter values in case of best fitness (as in Table 6.1). Δ*ff* = *ff* − best(*ff*)/best(*ff*), where Δ*ff* indicates the percentage change in fitness compared to the maximum fitness reported in Table 6.1 when changing the parameter by 10%. If a 10% change leads to a value of *th*, *pclm* or *pclt* >1, the value is set =1.

on the three-parameter configuration used to derive the three best-ranked fitness measures. Table 6.2 reports the sensitivity analysis results. A sensitivity analysis allowed us to assess the importance of single parameters. From Table 6.2, we can infer how the biggest changes in loss of fitness happened with respect to a change in the 'positive experience threshold', where an increase in the threshold by 10% (or 1 whichever is maximum) led to a loss of fitness of over 140%, while a reduction in the threshold by 10% led to a loss of fitness of over 60%. From the sensitivity analysis, the positive experience threshold was the most important parameter, followed by competitiveness. Also, competitiveness was the only parameter that, with respect to the configuration of the second-best fitness, led to an actual increase in fitness. This is possible since search algorithms approximate best solutions rather than finding them with certainty (see the discussion in Chapter 2 on calibration, fitness and sensitivity). On the other hand, loss of fitness due to cost differences seemed to be contained, all else being equal, as well as the weight given to changes in climate, culture and experience. Finally, the probability of an agent preferring culturally similar or diverse countries had the most 'differentiated effect'. In fact, decreasing the probability of preferring culturally similar countries had a much smaller effect than increasing it.

To conclude, our analysis showed that the main sensitivity of our model lies in the choice of the positive experience threshold parameter, and secondly in the probability of agents' preference to travel to culturally similar countries. One exception is the importance of an increase in cost differences (for the second-ranked model, where an increase in *wcst* caused a loss of fitness of 20.04%).

Predicting Cancellations of Hotel Bookings

Problem statement

Revenue management (RM) is an umbrella term for a set of strategies that enable capacity-constrained service industries to realise optimum revenue from operations. The core objective is to provide the right service to the right customer at the right time for the right price. It involves carefully defining service, customer, time and price, and is supported by a number of modelling and optimisation methods in deciding what to sell, when to sell, to whom to sell and for what price, in order to increase revenue and profit. For a hotel, RM is a complex framework characterised by many parameters and constraints, in which stochasticity, especially in the demand, complicates the situation. As for other areas, exact solutions are not always possible, and one has to resort to some intelligent heuristics and simulations. Effective methods today include learning techniques that update models based on the past history of the hotel considered and of the environment in which it is located (Talluri & Van Ryzin, 2006). One important task for RM, the object of this case, is predicting cancellations that can hamper accurate forecasts (Antonio *et al.*, 2017).

Methods: Choice and use

Although the term *prediction* is used, what we are really dealing with is a classification task. Given certain booking characteristics and past history, we want to know whether new bookings may or may not lead to cancellation. To do that, we use one of the suitable ML algorithms (see Chapter 3: Machine Learning). Possible candidates, among many, we can test are logistic regression, naïve Bayes and a support vector machine (SVM). Given that we have more than one potential technique appropriate for the problem statement at hand, we build the different models and check their accuracy by splitting the input data into a training set and a test set (these classification algorithms, as seen, are supervised algorithms). The model built on the training set is then applied to the test set and the resulting split compared to the actual values. The accuracy of the classification allows the selection of the most accurate method for our task.

The measurement of this accuracy is made by calculating the following quantities:

- *precision*: number of correct (true) positive results divided by the number of all positive results returned by the classifier (true and false positives);
- *recall*: number of correct positive results divided by the number of all samples that should have been identified as positive (true positive and false negatives);
- *F1 score*: the harmonic mean of the precision and recall. F1 varies between 1 (perfect precision and recall) and 0; and
- *accuracy*: number of correct predictions (true positives) divided by the total number of items classified.

These quantities are usually arranged in a matrix called a *confusion matrix* (or error matrix). For the calculations, we use Rapidminer (see Appendix 2).

Data collection and preparation

Data were from the hotel property management system (PMS), the software application used for the operational functions of a front office.

The actual data sets used here have been made freely available by Antonio *et al.* (2019).

The two data sets were downloaded (and anonymised) from the PMSs of two Portuguese establishments: a resort in the Algarve (H1) and a hotel in Lisbon (H2). Both data sets shared the same structure, with 31 variables describing the 40,060 observations for H1 and the 79,330 observations for H2. Each observation represented a hotel booking. All bookings were marked (labelled) as cancelled or non-cancelled, depending on their final result.

For our purposes, we first selected the 'meaningful' variables: lead time (days between reservation and arrival); reservation for weekend or weekdays; number of adults, children and babies; meals; country of origin; distribution channel type; returning guest; previous cancellations; previous bookings not cancelled; room type reserved and assigned; booking changes; deposit type; agent; company (for business travellers); days on waiting list; customer type (single, group, etc.); average daily rate; required car parking spaces; number of special requests.

We used H1 to build a model and then applied it to the second data set (H2) to check the overall validity. A quick analysis of H1 showed that there was an imbalance between the number of cancelled and non-cancelled bookings: almost 72% of reservations were satisfied. This situation was at risk of *overfitting* in a supervised (as in our case) classification task. In other words, we risked fitting the data *too well*, underestimating the cancellations and giving more (and excessive) weight to the positive bookings (Lever *et al.*, 2016). To solve this issue, we randomly selected the records and built a balanced data set containing the same number of cancelled and non-cancelled bookings (10,000 each in our case). To check the models derived from H1, we also randomly selected 20,000 records from the H2 data set.

The Rapidminer setting is shown in Figure 6.3. Once the data set was loaded, the Select Attributes operator (the term used for an algorithm or procedure) allowed the choice of the desired variables. The Validation ran the model and produced the outcomes. It is a 'nested' operator: it contains the model chosen (in Figure 6.3 it is a naïve Bayes) that is applied to the data set once it has been split into a training and testing part (here we opted for a training set including 75% of the records). The model was built using the training set and was then applied to the testing set. The performance operator provided the wanted metrics that allowed judgement of the classification quality. By simply changing the model (from naïve Bayes to SVM or logistic regression), we finally chose the one that best fit our data.

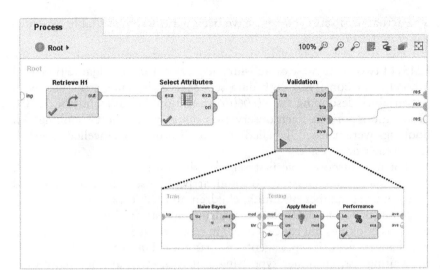

Figure 6.3 Rapidminer operators for the prediction of cancellations

Application of the models obtained to the second data set was performed in almost the same way. This time, however, the training set and the test set came from the two different data sets H1 and H2. The Rapidminer setting is shown in Figure 6.4; here too the naïve Bayes algorithms can be easily changed into one of the other models that best fits our data.

Finally, it must be noted that some operator implementations can have different requirements concerning the type of variables used. In this case, the software will advise and provide guidance.

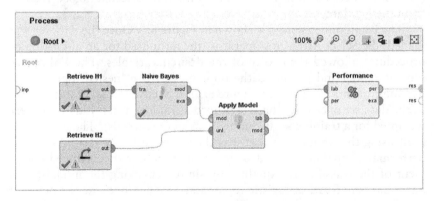

Figure 6.4 Rapidminer setting for the application of a model on H2

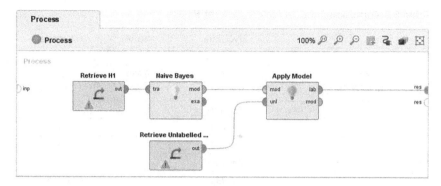

Figure 6.5 Prediction of cancellations

If we need to predict possible cancellations, we use a data set of bookings with the same format as the one used to build the model (H1 in our case) that does not contain the cancelled/non-cancelled label. The output will consist of a data set with the label applied to each reservation (Figure 6.5).

Worked example

The results of the study provide the software with the metrics for judging the accuracy and the validity of the models used, thereby allowing the selection of the best possible algorithm for the specific case and situation.

One consideration is in order here. The procedure described contains several 'random' choices (record selection, determination of training set, etc.), and the algorithms used are inherently stochastic in nature. The presence of random choices with stochastic algorithms risks producing results that might be classified as 'special cases'. To avoid the special cases issue, it is advisable to repeat the whole process N certain number of times and average the results, thus smoothing possible extreme values. As in all simulations, one run is only a special case of a specific parameter combination.

The results of the first series of calculations are shown in Table 6.3, which contains the confusion matrix for each classification algorithm employed. As can be seen, the models generally have good accuracy and the logistic regression is the best-performing method. The same models applied to the second data set (H2) provide the outcome shown in Table 6.4, and here too, the logistic regression performs best.

Table 6.3 Confusion matrices for the data set H1

	Precision	Recall	F1-score
Logistic regression	Accuracy:	0.886	
Non-cancelled	0.885	0.880	0.885
Cancelled	0.885	0.890	0.890
SVM	Accuracy:	0.774	
Non-cancelled	0.760	0.800	0.780
Cancelled	0.790	0.740	0.770
Naïve Bayes	Accuracy:	0.832	
Non-cancelled	0.935	0.695	0.775
Cancelled	0.795	0.970	0.865

Table 6.4 Confusion matrices for the data set H2

	Precision	Recall	F1-score
Logistic regression	Accuracy:	0.843	
Non-cancelled	0.890	0.825	0.855
Cancelled	0.805	0.865	0.830
SVM	Accuracy:	0.708	
Non-cancelled	0.730	0.790	0.760
Cancelled	0.670	0.590	0.630
Naïve Bayes	Accuracy:	0.733	
Non-cancelled	0.855	0.600	0.655
Cancelled	0.710	0.920	0.780

Table 6.5 Percentage difference between models for H1 and H2

	Precision	Recall	F1-score
Logistic regression	Accuracy:	5%	
Non-cancelled (%)	−1	6	3
Cancelled (%)	9	3	7
SVM	Accuracy:	8%	
Non-cancelled (%)	4	1	3
Cancelled (%)	15	20	18
Naïve Bayes	Accuracy:	12%	
Non-cancelled (%)	9	14	15
Cancelled (%)	11	5	10

Finally, we note that the same model is the one that has the lowest difference (%) between the two cases (Table 6.5), thus confirming its good applicability in these two cases, which probably depends on the similarity of the market and environmental conditions (the two establishments are in the same geographic area).

The Importance of Networking for a Hotel

Problem statement

Cooperation and collaboration are important issues, and have been widely discussed in many works that have highlighted the importance of a good and efficient set of relationships that can help a hotel achieve valuable results (Baggio, 2011; Bramwell & Lane, 2000). A network of such collaborations allows access to resources and information as well as increasing the attractiveness of a specific product by combining it with other local attractions (Sainaghi, 2006; Xiang & Pan, 2011).

Hotels in a tourism destination have connections with a number of organisations (travel agencies and tour operators, entertainment firms, destination management organisations, local associations, etc.) to increase their attractiveness to potential tourists. The question we are interested in here is whether and to what extent the characteristics of the set of relationships each hotel has can, in some way, be connected to their operational results. This example is based on the work of Sainaghi and Baggio (2014).

Methods: Choice and use

Since the main element in this discussion is a set of relationships, a network model is the natural choice. The idea is to build the whole network of a destination, to assign each hotel in the network a value that can represent the *quality* of its position in the network and to compare these values to some index of performance and analyse the results.

The destination used in this work is Livigno (Italy), a mature mountain destination located in northern Italy, close to the Swiss border, with a well-established identity and a rich offering of products and services for its visitors. In terms of operating tourism stakeholders, Livigno has about 1000 companies. Livigno receives about half a million visitors per year, with a strong seasonality.

Once the network was built, we calculated the basic normalised nodal-level metrics (see Chapter 3: Network Analysis): degree, closeness, betweenness, local efficiency, clustering coefficient and eigenvector centrality. Each centrality metric implies some kind of 'importance' of the node; therefore, we combined all these metrics into a single index using the geometric mean of the values (normalised values are ratios, and a geometric mean is less sensitive to extreme values). We then compared these values with the hotels' occupancy, one of the main metrics used for assessing a hotel's performance.

Data collection and preparation

The network was built following the methods discussed in Chapter 3: Network Analysis. After a list of all stakeholders had been compiled, the links between them were enumerated looking at historical records and websites' hyperlinks. These data were complemented with interviews with the most relevant actors. The occupancy data came from one of the authors' ongoing records and were averaged over the previous three years.

Worked example

Once all the needed quantities had been measured, the simplest way to proceed was to calculate the correlation existing between the two variables: NET, the network position quality and occupancy. The results are shown in Table 6.6 in which both the Pearson and Spearman coefficients are reported, and in Figure 6.6.

As can be seen from Figure 6.6, the correlation was good but it would have been better if we could have removed the three outliers marked with a star (here, a good knowledge of the specific situations might have provided good suggestions).

However, to better understand the significance of these results, we needed a null model, a reference setting that we could use to compare the outcomes. To build a null model, following common practice when dealing with networks, one generally randomises the network by rewiring all the links but retaining unchanged both the number of nodes and the degree of each node (see e.g. Squartini *et al.*, 2011; Squartini & Garlaschelli, 2011). Following randomisation, Spearman's correlation was calculated again, this time assuming a value of 0.267 (significant at the 0.05 level). This much lower value confirmed the relevance of the results obtained.

The last question that could be asked is whether there was a causal relationship between the two quantities. Usually, a correlation is not

Table 6.6 Correlation coefficients

Pearson's correlation	0.443 (sig. two-tailed: $<10^{-4}$)
Spearman's rho	0.452 (sig. two-tailed: $<10^{-5}$)

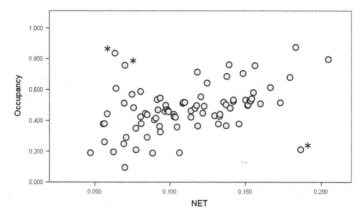

Figure 6.6 Correlation between hotels' occupancy and quality of network position (stars indicate three outliers that could be removed) (adapted from Sainaghi & Baggio, 2014)

considered a sufficient reason to imply causation, unless at least two more conditions are satisfied (Cook & Campbell, 1976; Hatfield *et al.*, 2006): the cause temporally precedes the effect and alternative explanations for a possible link between cause and effect can be ruled out.

In our case, all these conditions seemed to be met: we had a good correlation, and other explanations for obtaining good results could be dismissed. In fact, as shown in Sainaghi and Baggio (2014), correlation between occupancy and location or size or category (usually possible determinants) was not significant. Lastly, assuming reverse causation was highly improbable. From a logical point of view, it is difficult to think that having a high occupancy would push a hotelier to establish more relationships with other organisations (the opposite is probably true), and the building of relationships is either independent from any performance outcome or networks are normally *older* than the recording of occupancies used in this case.

Tourism Development and the Environment: A Long-Term Perspective

Problem statement

In some communities the tourism sector is often the main provider of economic opportunities and well-being. Tourism has expanded considerably in terms of growth in infrastructure and facilities in order to facilitate travel, increasing the attractiveness of a tourism destination by building amenities and accommodation offerings. However, the enduring growth of the tourism sector may impact the environment. Long term, this can be detrimental, especially for tourism based on natural resources (beaches, sun, pristine environments).

In their work, Casagrandi and Rinaldi (2002) were concerned with the relationship between tourism development and environmental sustainability. In fact, they contended that it may not be possible to devise policies that simultaneously achieve two key objectives: long-term maintenance or growth of the tourism sector and environmental sustainability (in their words, avoid severe impacts on the environment). Thus, the problem statement intrinsically refers to the long-term dynamics and the coupling of the tourism system with the environment. The example reported is based on the work of Casagrandi and Rinaldi (2002).

Methods: Choice and use

Given the importance of long-term dynamics for the problem statement at hand, as well as the reduced data availability that proxy tourism development and environmental conditions in a specific destination, coupled with the intent to be general, and avoid contextual characteristics that may affect the outcome of such study, the authors employed a stylised model (also called a minimal model). While very abstract, stylised models can shed light on important dynamics that unfold in complex systems. Given the abstraction of such models, they are simple to understand and allow for a clear comprehension of the effects of specific parameters and variables on the outcome. In other words, such models are characterised by the inclusion of very limited, abstract but at the same time meaningful variables or parameters that can impact the outcome of interest. In this case, a system dynamic model was the model of choice. The reasons were the relative simplicity of such models, the interest in an aggregated outcome coupled with the assessment of long-term dynamics, and the generality of the model needed to answer or assess the problem statement at hand. It is important to reiterate that the problem statement does not imply the differentiation of different types of tourists and preferences (as in the previous example), hence a system dynamic approach is more conducive to matching the problem statement than an ABM.

Data collection and preparation

Given the nature of the model, and the methods of choice, data were not needed. This is a theoretical model for which parameters and variables are mainly used. Such parameters are thus purely theoretical and

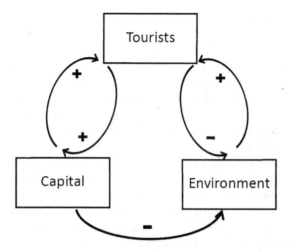

Figure 6.7 Schematic representation of the system dynamic model

are chosen to represent the dynamics that the model seeks to explain. The variables are chosen to represent key characteristics that are important for the problem statement; in this case: Tourists (T), Environment (E) and Capital investment in infrastructure (C). The elements of the model represented core features common to many tourism systems. In short, no data were collected and parameters and variables will be used in the model as portrayed in Figure 6.7.

Worked example

Following Casagrandi and Rinaldi (2002), it was important to assess how the three main components of the model interacted. In the example used here, tourists could communicate the attractiveness of the destination visited. Further, this attractiveness influenced future tourist choice to visit (or not) that specific destination. Attractiveness was then assumed to be relative. That is, tourists evaluated the attractiveness of a tourist destination differently, but always in relation to an 'absolute attractiveness' that was due to tourism, infrastructure and the quality of the environment. Although attractiveness may depend on variables such as cost and the relative attractiveness of other destinations, it also depends on the environment and the infrastructure. Casagrandi and Rinaldi (2002) thus modelled attractiveness as a function of E and C. Further, it was assumed that the Environment was not a rivalrous good (i.e. one's enjoyment of the environment was not reduced by the number of tourists enjoying it), while infrastructure was a rivalrous good. In our opinion, this assumption was quite simplistic and did not take into account the type of tourists and the crowding effect at tourists' sites related to crowded beaches, national parks and so on. However, it was a reasonable assumption considering a specific number of tourists. However, the crowding effect (congestion in Casagrandi and Rinaldi's words) was considered overall, that is, while E was not directly dependent on tourists, the attractiveness of the destination linearly decreased with the number of tourists in a specific destination.

In this stylised model, the environment was dependent on the quality of a location's environment without tourism (but included existing infrastructure for industry, services and other activities), and deteriorated depending on the damage caused by tourism. Finally, capital was dependent on the investment levels in infrastructure. See Casagrandi and Rinaldi (2002) for detailed information of the model and the model equations used, the choice of functional forms (or how components were related) and the actual parameters used in the model.

Further, parameters affecting attractiveness, tourist numbers, infrastructure and the environment were subdivided into policy and system parameters. Parameters that affected the behaviour of tourists as well as decision makers (i.e. decision to invest in infrastructure) were thought of as policy parameters, while parameters that affected the environment as well as the attractiveness of an alternative destination were thought of as system parameters. Hence, changing parameter values (parameters are constant during a simulation run) was akin to assessing a specific policy or system feature of the model and how this impacted the overall results on tourist flow, infrastructure development and the quality of the environment. The model can be implemented using the system dynamic modeler available in the NetLogo environment (see Figure 6.8).

Running the model under different parameter configurations showed the combined effects of the three main components and the consequences of different possible policies. An important outcome was that policies that aimed to enhance tourism development were not able to be combined with a reduction or a low impact on the environment. As Casagrandi and Rinaldi (2002: 15) state: 'Safe policies are "surrounded" by risky policies, and a continuous increase in the system parameter transforms a safe sustainable policy first into a risky sustainable policy and then into an unsustainable policy. In general, sustainability requires low prices, low investments, and high environmental reclamation'.

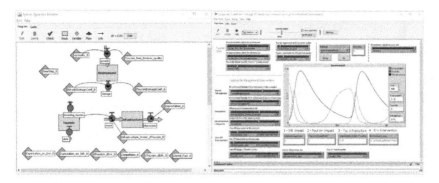

Figure 6.8 NetLogo system dynamic build and model interface implementation (adapted from the version by Boschetti at www.per.marine.csiro.au/staff/Fabio.Bosc hetti/ToyModels/ToyModels.htm)

Concluding Remarks

In this final chapter, we have discussed four different cases related to tourism and hospitality in which the methods and techniques discussed in this book can or have been applied. The objective of these examples is to show the reader, at a very general level, the reasoning behind specific modelling choices that mainly depend on the problem statement we are interested in and data constraints/availability (as explained in Chapter 5). Further, these examples also showcase the type of outcomes that can be obtained. Such outcomes may vary and need further analysis. Especially when it comes to simulated outcomes, it is necessary to portray some aggregated measure of multiple simulation runs, as portraying only one outcome of one simulation can and in almost all cases does lead to wrong conclusions. The focus of this chapter was thus on the modelling choice and the implementation of specific modelling methods as well as portraying the results of such methods. Thus, we have skipped the usual comments on limitations or on how to use the results for informing actions or making decisions, limiting ourselves to a simple description of the problem at hand, a basic motivation for the choice of modelling tools and outcomes for the case examined. The reader will surely be able to make more considerations and guess the possibilities, the limitations and the advantages these applications may have in a real-life environment.

References

Amelung, B. and Moreno, A. (2009) Impacts of Climate Change in Tourism in Europe. PESETA Tourism Study. JRC Scientific and Technical Reports EUR 24114. Publications Office of the European Union, Luxembourg.

Antonio, N., de Almeida, A. and Nunes, L. (2017) Predicting hotel booking cancellations to decrease uncertainty and increase revenue. *Tourism & Management Studies* 13 (2), 25–39.

Antonio, N., de Almeida, A. and Nunes, L. (2019) Hotel booking demand datasets. *Data in Brief* 22, 41–49.

Baggio, R. (2011) Collaboration and cooperation in a tourism destination: A network science approach. *Current Issues in Tourism* 14 (2), 183–189.

Bigano, A., Hamilton, J.M. and Tol, R.S. (2006) The impact of climate on holiday destination choice. *Climatic Change* 76 (3–4), 389–406.

Bramwell, B. and Lane, B. (2000) *Tourism Collaboration and Partnerships: Politics Practice and Sustainability*. Clevedon: Channel View Publications.

Casagrandi, R. and Rinaldi, S. (2002) A theoretical approach to tourism sustainability. *Conservation Ecology* 6 (1), art. 13.

Cohen, A.B. (2009) Many forms of culture. *American Psychologist* 64 (3), 194–204.

Cook, T.D. and Campbell, D.T. (1976) The design of quasi-experiments and true experiments in field settings. In M.D. Dunnette (ed.) *Handbook of Industrial and Organizational Psychology* (pp. 223–326). Chicago, IL: Rand McNally.

Eurostat (2016) Tourism statistics. See http://ec.europa.eu/eurostat/statistics-explained/index.php/Tourism_statistics (accessed April 2017).

Gursoy, D. and McCleary, K.W. (2004) Travelers' prior knowledge and its impact on their information search behavior. *Journal of Hospitality & Tourism Research* 28 (1), 66–94.

Hatfield, J., Faunce, G.J. and Job, R.S. (2006) Avoiding confusion surrounding the phrase 'correlation does not imply causation'. *American Psychologist* 33 (1), 49–51.

Hofstede, G. (2001) *Culture's Consequences: Comparing Values, Behaviors, Institutions and Organizations across Nations*. London: Sage.

Lehto, X.Y., O'Leary, J.T. and Morrison, A.M. (2004) The effect of prior experience on vacation behavior. *Annals of Tourism Research* 31 (4), 801–818.

Lever, J., Krzywinski, M. and Altman, N. (2016) Points of significance: Model selection and overfitting. *Nature Methods* 13 (9), 703–704.

Numbeo (2016) Cost of living. See https://www.numbeo.com/cost-of-living/ (accessed April 2017).

Oppermann, M. (2000) Tourism destination loyalty. *Journal of Travel Research* 39 (1), 78–84.

Sainaghi, R. (2006) From contents to processes: Versus a dynamic destination management model (DDMM). *Tourism Management* 27 (5), 1053–1063.

Sainaghi, R. and Baggio, R. (2014) Structural social capital and hotel performance: Is there a link? *International Journal of Hospitality Management* 37, 99–110.

Scott, D., Rutty, M., Amelung, B. and Tang, M. (2016) An inter-comparison of the holiday climate index (HCI) and the tourism climate index (TCI) in Europe. *Atmosphere* 7 (6), 80–97.

Squartini, T. and Garlaschelli, D. (2011) Analytical maximum-likelihood method to detect patterns in real networks. *New Journal of Physics* 13 (8), art. 083001.

Squartini, T., Fagiolo, G. and Garlaschelli, D. (2011) Randomizing world trade. I. A binary network analysis. *Physical Review E* 84 (4), art. 046117.

Talluri, K.T. and Van Ryzin, G.J. (2006) *The Theory and Practice of Revenue Management*. New York: Springer Science & Business Media.

UNWTO (2014) *Tourism Statistics: Intra-EU Tourism Flows*. Madrid: World Tourism Organization.

UNWTO (2016) *Tourism Highlights, 2016 Edition.* Madrid: World Tourism Organization.

WEF (2015) *The Travel & Tourism Competitiveness Report 2015.* Geneva: World Economic Forum.

Williams, A.M. and Baláž, V. (2015) Tourism risk and uncertainty: Theoretical reflections. *Journal of Travel Research* 54 (3), 271–287.

Woodside, A.G. and Lysonski, S. (1989) A general model of traveler destination choice. *Journal of Travel Research* 27 (4), 8–14.

WTTC (2016) *Travel & Tourism Economic Impact 2016 World Report.* London: World Travel & Tourism Council.

Xiang, Z. and Pan, B. (2011) Travel queries on cities in the United States: Implications for search engine marketing for tourist destinations. *Tourism Management* 32, 88–97.

A Closing Remark

Not many years ago, the study of complex phenomena and systems, such as the domain of travel and tourism, was greatly hindered by the necessity to use highly simplified representations and schematic methodologies. In the last decade, the notable progress in, and the wider accessibility of, a large set of computerised tools, together with the availability of suitably powerful hardware systems, has made an approach based on the design and implementation of numerical modelling and simulation environments practicable.

Many of these methods are also supported and conditioned by a huge influx of data, mostly coming from the virtual world we spend time in, leading to the necessity to rethink many of the methods that have been developed for analysing and extracting information.

Simulation and modelling tools and techniques are of great importance today, as they provide the means for better understanding the complexity of real processes, systems and phenomena. They also allow ways of experimenting in situations where this important practice is not workable for the type of objects we want to study, as in the case of socioeconomic domains, or for the number and the type of phenomena involved or for the time scale of the 'natural' developments we are interested in. It is widely recognised that well designed and carefully conducted experiments have the same power and validity as their impossible 'real' equivalents, thus allowing the possibility to create reliable scenarios or predictions (where feasible) that are of great interest from both a theoretical and a practical perspective where they greatly support making better-informed decisions.

Often, computer simulation and modelling programs are not comparable to any other computer program. Instead of providing a fixed answer to a precise question, they provide dynamic environments in which some synthetic representation of a system is actioned based on certain algorithms. Thus, they are particularly sensitive to the choices a modeler makes in terms of the time scales adopted and the variables used for describing the different systems' features. Hence the importance of good calibration, verification and validation methods and practices.

Another distinctive characteristic of this domain is its overcoming the distinction between qualitative and quantitative approaches to the solution of a problem or the understanding of a phenomenon. The two, often artificially competing, attitudes need to be complemented if good and useful outcomes are to be obtained. A famous quote attributed to William Deming, American engineer and management consultant and considered father of that 'total quality' that greatly influenced Japanese industrial development in the second half of last century, is: 'Without data, you're just another person with an opinion'. In the modelling and simulation domain, however, this should be accompanied by a complementary: 'without an opinion you're just another person with data', as the interpretation, soundly based on qualitative knowledge of what we do and obtain, plays a crucial role. Both quantitative and qualitative instruments are necessary to fully exploit the potential of the methods presented in this work, not only for practical purposes, but also because as Gummesson (2007: 226), for example, recognises: 'By abolishing the unfortunate categories of qualitative/quantitative and natural sciences/ social sciences that have been set against each other, and letting them join forces for a common goal – to learn about life – people open up for methodological creativity, therefore qualitative and quantitative, natural and social are not in conflict but they should be treated in symbiosis'.

The importance of treating the natural and social sciences as symbiotic has long been known. Greek philosophers such as Aristotle and the Stoics clearly stated that real systems should and could be analysed and understood only as the synthesis of four key aspects (see also Nijland, 2002):

(1) Physics: Observed or perceivable quantities.
(2) Logics: Theories (e.g. mathematical constructions or computational models).
(3) Ethics: Norms and values in which the system we are studying is embedded (e.g. the environmental context).
(4) Politics: The actions and interactions that shape and are shaped by the phenomena we are studying.

Hence, understanding complex systems, and tourism systems more specifically, requires an understanding of observed quantities, interactions, norms and values. A tourism system cannot be 'understood' just by examining feedback between deductive reasoning and empirical testing (Figure c.1A), it also requires assessing and analysing the interactions and actions (*politics*) and the model of values (*ethics*) (Figure c.1B).

For example, as van Gigch (2002) states, management science has tried to formalise the decision-making processes, often making narrow assumptions on the model of values (e.g. greed) and on the influences of actions, thus not looking at action and values to take into account clients–recipients relations. However, as Nijland (2002: 212) points out, an unarticulated holistic approach is not useful; 'a longitudinal, interdisciplinary, participatory but quantitative social-system analysis' is preferred. Hence, once again, in these concluding remarks, we recommend researchers interested in analysing, assessing, explaining and predicting tourism systems to be open to multiple methodological approaches as well as to acquiring knowledge and experience within an interdisciplinary frame of reference. If our aim is to increase our understanding of a tourism system, no matter which aspect is prevalent and which methodology we are more familiar, we will only be able to make meaningful advances by combining the four aspects described over 2000 years ago by Aristotle. Hence, we invite researchers to adopt an approach such as that depicted in Figure c.2 in an interactive and non-consecutive manner. Exploiting different knowledge domains (ethics, politics, deductive reasoning and empirical testing) also requires a multi-methodological approach in which one recursively uses case studies (interviews, surveys, direct observation), modelling (computational and analytical) and experiments (Figure c.2).

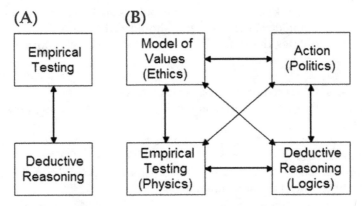

Figure c.1 Interrelation between cognitive acts without (A) and with (B) action and model of values; arrows indicate the relations between the 'knowledge domains'; the name given by Aristoteles to the different knowledge domains is in parentheses

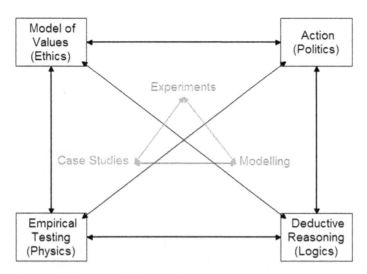

Figure c.2 A possible approach for studying tourism systems

Given what we have discussed in this book, it is possible to employ modelling techniques to aid cognition and make sense of the observed reality, as well as to represent and simplify both the model of values (or context) and the effects of future and past actions and how these interact with the observed reality. In other words, modelling can aid in defining and focusing cases and experiments, and both case studies and experiments are at the basis of the 'expert' knowledge needed to develop rigorous modelling assumptions.

Since tourism systems are open systems, problems of replicability or applicability, as we have seen throughout this book, will definitely arise. Nonetheless, such issues need to be brought forth and if possible addressed, once a first representation of reality (model) that is logical and consistent, as well as rigorous, is built and put in place. Once again, it is worth emphasising that often the key purpose of modelling in tourism may not be limited to predicting the behaviour of single actors (e.g. individuals), inherently very difficult if not impossible to predict, but should also examine the behaviour (explaining or predicting) of the overall system in a specific area, as an aggregate. Reaching such an explanation, as we have seen, will require employing different modelling techniques, in sequence or mixing them up, depending on the problem statement we want to address as well as on the availability of suitable data and other potential constraints. Thus, modelling tourism systems should be devoted not only (and often not at all) to predicting the future behaviours of the system itself, but also to understanding if and how regularities may emerge (Majorana, 1942), eventually trying to generate and assess reasonable scenarios of the overall system in order to be able to intervene and incentivise potential desirable settings.

This methodological creativity, based also on a valuable multidisciplinary approach, is well represented in the world of modelling and simulation. A relatively simple and concentrated description of the main methodological issues and the most relevant techniques for modelling and simulating is the objective of this work. The aim is to increase the awareness of tourism and hospitality researchers and practitioners in these practices to better account for the complexity of the phenomenon and of the systems that are included in the common notion of the domain. Moreover, the adoption of good research practices that take into consideration the complexity characteristics of the tourism domain and of more rigorous, multiform and flexible methodological tools such as those described in this work will aid research in moving towards a more disciplined array of theories and models and contribute to a better methodological foundation for tourism (Tribe, 1997).

Much more needs to be examined, studied and attempted for those who want or need to successfully use these methods. In this book, we have only provided a 'first aid' kit that may better drive interested parties to the complicated but fascinating art of modelling and simulation.

References

Gummesson, E. (2007) Case study research and network theory: Birds of a feather. *Qualitative Research in Organizations and Management* 2, 226–248.

Majorana, E. (1942) Il valore delle leggi statistiche nella fisica e nelle scienze sociali. *Scientia* 71, 58–66.

Nijland, G.O. (2002) The tetrahedron of knowledge acquisition: A meta-model of the relations among observation, conceptualization, evaluation and action in the research on socio-ecological systems. *Systems Research and Behavioral Science* 19 (3), 211–221.

Tribe, J. (1997) The indiscipline of tourism. *Annals of Tourism Research* 24, 638–657.

van Gigch, J.P. (2002) Comparing the epistemologies of scientific disciplines in two distinct domains: modern physics versus social sciences. II: Epistemology and knowledge characteristics of the 'new' social sciences. *Systems Research and Behavioral Science* 19 (6), 551–562.

Appendix 1: Further Readings

Appendix 1 contains a list of books on the different topics discussed in this book. With absolutely no pretence of completeness, the list is just a few suggestions and starting points for readers who may want additional and more detailed coverage of the various matters. Many of these books are also complemented by online supplemental materials such as data and software scripts. An online search will help find these useful additions.

Conceptual Modelling

Very often the theme is treated as a starting activity in works on research methods or qualitative analysis books. A good reference is

- Embley, D.W. and Thalheim, B. (eds) (2012) *Handbook of Conceptual Modeling: Theory, Practice, and Research Challenges*. Heidelberg: Springer Science & Business Media.

Although loosely connected, mind mapping is a good method for graphically structuring ideas and concepts. The basic method is treated in

- Buzan, T. (2006) *The Ultimate Book of Mind Maps*. London: Harper Thorsons.

Not specifically on conceptual modelling, but with several examples that can be classified as such is

- Pearce, D. (2012) *Frameworks for Tourism Research*. Wallingford: CABI.

Statistical Modelling

The literature on statistical methods is quite vast and includes many books covering all aspects at any level, from the simplest descriptions to the most complicated and sophisticated treatments.

A general reference, with specific examples implemented in SPSS, one of the most used programs in the field is

- Field, A. (2017) *Discovering Statistics Using IBM SPSS* (5th edn). London: Sage.

Specifically targeted at the tourism domain are

- Baggio, R. and Klobas, J. (2017) *Quantitative Methods in Tourism: A Handbook* (2nd edn). Bristol: Channel View Publications.
- Smith, S.L.J. (2013) *Tourism Analysis: A Handbook* (2nd edn). New York: Routledge.

Machine Learning and Artificial Intelligence

The recent great interest in this area has led to a wealth of publications from popular introductions to highly sophisticated treatments that require well-developed skills. Many good introductory texts are, however, relatively old. Starting points of a generally average level are

- Bishop, C.M. (2006) *Pattern Recognition and Machine Learning*. New York: Springer-Verlag.
- Hastie, T. and Tibshirani, R. (2017) *The Elements of Statistical Learning: Data Mining, Inference, and Prediction* (2nd edn). New York: Springer.
- Mitchell, T.M. (1997) *Machine Learning*. New York: McGraw-Hill.
- Russell, S.J. and Norvig, P. (2016) *Artificial Intelligence: A Modern Approach* (3rd edn). London: Pearson.

System Dynamics Modelling

A good and well-cited introduction is

- Meadows, D. (2008) *Thinking in Systems: A Primer* (ed. D. Wright). River Junction, VT: Chelsea Green.

Further helpful references are

- Coyle, R.G. (1996) *System Dynamics Modelling: A Practical Approach*. London: Chapman & Hall.
- Sherwood, D. (2002) *Seeing the Forest for the Trees: A Manager's Guide to Applying Systems Thinking*. Boston, MA: Nicholas Brealey International.
- Sterman, J.D. (2000) *Business Dynamics: Systems Thinking and Modeling for a Complex World*. Boston, MA: McGraw-Hill.

Network Science

This area, too, has seen big developments in recent years, and many texts, technical and popular, are available. The most complete, although not too technically advanced, references are

- Barabási, A.L. (2016) *Network Science*. Cambridge: Cambridge University Press.
- Newman, M.E.J. (2010) *Networks: An Introduction*. Oxford: Oxford University Press.

along with the classical

- Wasserman, S. and Faust, K. (1994) *Social Network Analysis. Methods and Applications*. Cambridge, MA: Cambridge University Press.

Specifically dedicated to the tourism domain:

- Scott, N., Baggio, R. and Cooper, C. (2008) *Network Analysis and Tourism: From Theory to Practice*. Clevedon: Channel View Publications.

Agent-based Modelling

Clear, comprehensive and accessible accounts can be found in

- Miller, J.H. and Page, S.E. (2007) *Complex Adaptive Systems: An Introduction to Computational Models of Social Life*. Princeton, NJ: Princeton University Press.
- Railsback, S.F. and Grimm, V. (2012) *Agent-based and Individual-based Modeling: A Practical Introduction*. Princeton, NJ: Princeton University Press.

A complete reference with many examples implemented in NetLogo is

- Wilensky, U. and Rand, W. (2015) *An Introduction to Agent-based Modeling: Modeling Natural, Social, and Engineered Complex Systems with NetLogo*. Cambridge, MA: MIT Press.

Appendix 2: Software Programs

NB: All links provided here were current at the time of writing (2019). The reader is advised to check the validity of the link, as its volatility may be high. In all cases, an online search using the relevant words and the name of the resource as keywords will allow the materials desired to be located with relative ease.

Modelling and simulation activities can only be performed with the help of computer programs. Many programs are available today, with different degrees of sophistication and covering different sets of functions. Several commercial packages are well designed and stable, but these are usually quite expensive; however, many limited or discounted versions are available for academic purposes (both teaching and research). Additionally, there are many high-quality, free (or very low cost) programs.

In either case, the user community has produced and made available a wealth of examples, applications and tutorials. A simple web query containing a few keywords and the name of the software product will provide all the information and advice desired.

A distinction needs to be made between packages and development environments. The first category includes applications usually endowed with good graphical interfaces giving access to the many functions and parameters needed. Its basic usage is relatively simple, but its many functions and options may make for a rather steep learning curve. Moreover, it may use different conventions, definitions and algorithms so that results, at times, may be difficult to compare. No single product is able to satisfy all needs. Therefore, before adopting one of these packages, it is advisable to collect information (an online search will provide enough information) and possibly download a trial version (often available for commercial packages) and do a field test.

It is worth mentioning here that, at least for the simplest cases, 'multipurpose' applications such as a spreadsheet program (e.g. Microsoft Excel, http://office.microsoft.com/excel/ or the free OpenOffice, http://www.openoffice.org/) can be a useful tool. Their disadvantages are that they generally have difficulty accommodating large sets of data and, apart from some elementary functions, more complex procedures may require longer preparation and a good knowledge of the software's functions.

The second category is basically made up of programming languages equipped with user interfaces and specialised editors, which simplifies the development and testing of a script.

The issue here can be the learning phase of the language since writing computer code is not an easy task, but they offer the greatest possible flexibility for designing algorithms and procedures. In many cases, suitable libraries exist that perform the most common functions.

Programming Languages and Development Environments

Any programming language such as C++, Visual Basic, Java or Fortran is suitable, but some are generally considered to be more user-friendly than others and have become quite popular. Today, they are considered standard for all activities related to the domain of modelling and simulation.

Python and R are in this class and are considered to be the environments of choice for the modern researcher. They are completely free, platform independent and well supported by an incredible number of very sophisticated libraries. Moreover, since many journals today ask for papers to be supplemented with the data and the programs used, the reader will find that the vast majority use these languages. Python is at http://www.python.org/ and R is at http://www.r-project.org/.

While R originates from the statistical community, and thus is more 'equipped' for this type of analyses, Python was born as a general-purpose language, with much richer capabilities and more sophisticated treatment of data structures and computational techniques (distributed and parallel computing, for example).

One big advantage of using these environments is that it is possible to build complete models (or simulations) that 'smoothly' include different methods: analyse a network, calculate some statistical property and use a machine learning algorithm, for example. Most of the examples described in Chapter 3 could be handled by a single Python script.

A major difficulty with these environments can be the installation of the libraries required (because of their many prerequisites and compatibilities). The task can be discouraging. The reader is thus advised to resort to one of the available Python 'distributions' such as Anaconda (www.anaconda.com), Enthought Canopy (www.enthought.com/product/canopy/) or ActivePython (www.activestate.com/activepython).

These are packaged sets of libraries, preconfigured, harmonised and especially designed for use by researchers of any discipline and equipped with good integrated development environments (IDE) for scientific programming. One advantage of Anaconda over the others is that it also contains R together with libraries that link the two environments (Python and R), thus providing a complete toolset.

Other possible development environments are

Program	URL	Type
GAUSS	http://www.aptech.com/	Commercial
Ox/OxEdit (Gauss clone)	http://www.doornik.com/	Free
MATLAB	http://www.mathworks.com/	Commercial
Scilab (MATLAB clone)	http://www.scilab.org/	Free
Octave (MATLAB clone)	http://www.gnu.org/software/octave/	Free

It must be noted here that the clones cited show good, but not complete compatibility with the original language. The web provides good information about how to translate the scripts between them. Obviously, in many cases and for special functions, a direct translation is not possible, and the programs must be completely rewritten. Finally, clones can be a little less stable or reliable than their original counterparts.

Software Packages

The following list, by no means exhaustive, contains the most important and diffused tools divided according to the main modelling methods described in Chapter 3. Usually these programs are available for all operating systems and can run on existing computer hardware configurations, although at times they can be memory hungry, particularly when large data sets are involved. In any case, the reader is advised to check the installation prerequisites. All the websites cited here contain documentation, reference books, examples, tutorials and links to the different user communities.

Conceptual models

No software exists for conceptual models. The only support possible is provided by some graphical applications. Examples are the concept mapping and mind mapping software packages used to create diagrams of relationships between concepts, ideas or other pieces of information, which some consider more efficient than simply drawing on a piece of paper or a screen. An updated list can be found at en.wikipedia.org/wiki/

List_of_concept-_and_mind-mapping_software. Here, too, both commercial and free applications exist. One further possible advantage in using these tools is the capability, provided by some of them, to be used online for real-time collaborative work within a group.

Statistical models

These applications offer rich sets of functions and procedures, facilities to import and export data in different formats and interactive components for the generation of graphical representations. Some also have an internal programming language (but often too peculiar) with which the user can automate procedures, modify or extend the options available and create new analyses.

Program	URL	Type
SPSS	http://www.spss.com/	Commercial
MINITAB	http://www.minitab.com/	Commercial
SAS	http://www.sas.com/	Commercial
STATA	http://www.stata.com/	Commercial

Additionally, there are a large number of specific applications for general use that cover single techniques. The Wikipedia page en.wikipedia.org/wiki/List_of_statistical_packages provides a considerable list.

System dynamic models

The website of the System Dynamic Society contains a long list of tools for modelling and supporting all activities in this area (www.systemdynamics.org/tools). The most commonly used software packages are

- iThink and STELLA (www.iseesystems.com) commercial and educational licenses are available. The programs run on Windows and Mac computers.
- Powersim (www.powersim.com) available in several configurations, runs under Windows; different licensing schemes are proposed.
- Vensim (www.vensim.com) and Ventity (ventity.biz) are available for both Windows and Mac. Here, too, commercial and academic licenses are available.

An interesting Python library (PySD: github.com/JamesPHoughton/pysd) allows reading Vensim-designed models and running them in the Python environment. In this way, a system dynamic modeler can use some of the many computational tools developed in the larger data science community.

Agent-based models

Software applications in this area are basically a visual interface that shows the dynamic behaviour of the agents and the outcomes of the run (numerically or graphically) and a programming language for detailing the different agents' characteristics and their interactions.

The most popular is NetLogo (ccl.northwestern.edu/netlogo/), a free multi-agent programmable modelling environment. It uses a proprietary language, not excessively difficult to learn. The only drawback is that it can be very slow for large models or for exploring large parameter spaces.

Other known packages are

- AnyLogic (www.anylogic.com), a commercial package based on Java.
- Mason (cs.gmu.edu/~eclab/projects/mason/), java based with a free academic license.
- Repast (repast.github.io), free that can use different languages including Java, C++ and Python.

NetLogo and AnyLogic can also be used for system dynamic modelling.

Network models

The most frequently used packages in this area are

- UCINET (www.analytictech.com/), a commercial (although very cheap) program developed for Windows only.
- Pajek (http://mrvar.fdv.uni-lj.si/pajek/), free, Windows only (can run on Mac or Linux via Wine).
- Gephi (gephi.org), free, platform independent.

All these packages provide functions for a basic analysis, they also have little (or limited) support when special network features are involved (bipartite networks, link weights, directionality of links, etc.). More advanced methods or full treatment of special cases, as well as dynamic simulations, need to be addressed using the good and efficient libraries in the Python (NetworkX, igraph), R (igraph, sna, tnet etc.) or MATLAB environments. More packages and libraries are listed at en.wikipedia.org/wiki/Social_network_analysis_software.

Exponential random graph models (ERGM) are handled by

- PNet at www.melnet.org.au/pnet, a Java-based free software package.
- Statnet, a suite of R packages for network analysis based on ERGM models; an R library freely available at: www.statnet.org.
- ERGM at github.com/jcatw/ergm, a free Python implementation.

For the analysis of multilayer and multiplex networks, the following libraries are freely available:

- Muxviz, muxviz.net, based on R.

and the Python-based libraries:

- multiNetX at: github.com/nkoub/multinetx.
- Py3plex at: github.com/SkBlaz/Py3plex.
- Pymnet at: www.mkivela.com/pymnet/.

Machine learning

For personal use, probably the best offer available are the free Weka (www.cs.waikato.ac.nz/ml/weka/) and the commercial Rapidminer (rapidminer.com). They provide an effective and efficient user interface that offers a visual workflow designer with which the user can build procedures without being forced to write program code. In any case, they can interface Python programs if needed. Although commercial, Rapidminer offers an academic free license.

A mixture of open source, free and commercial packages for deep learning modelling can be found at en.wikipedia.org/wiki/Comparison _of_deep-learning_software.

For the most complex and computationally intensive tasks, all the major software providers have implemented cloud-based commercial web services that offer technical infrastructure and distributed computing tools with which analyses can be run or new applications developed. The major commercial services are

- Amazon Web Services (aws.amazon.com).
- Google Cloud (cloud.google.com).
- IBM Watson (www.ibm.com/watson).
- Microsoft Azure (azure.microsoft.com).
- Oracle Cloud (cloud.oracle.com).

All offer several functions priced on a pay-per-use basis, as well as consultancy, support and some free trials.

Beginners' Corner

For readers new to the world of modelling and simulations who wish to approach the domain, probably the best suggestion is to learn computer programming with Python and R. The Python.org website (https://www.python.org/) contains all the official documentation, a free online tutorial and pointers to a wealth of other material. The same can be said of CRAN (https://cran.r-project.org/), a network of ftp and web servers around the world that store up-to-date versions of code and documentation for R.

A useful platform is Codecademy (https://www.codecademy.com/). It offers free courses on many programming languages and provides free code editors so that users can practice. Courses are mostly free, and a low-cost upgrade allows exploration of more materials and examples, extra practice, advanced features and collaborate or share resources with the Codecademy community.

For those not willing to write code, a few packages can build a good toolbox.

Rapidminer Studio is a good choice for all statistical and machine learning analyses. The RapidMiner Academy (https://academy.rapidminer.com/) is well equipped with video tutorials on many functionalities of the software as well as explaining many data science concepts. The reader will also find a good selection of applications and use cases and even a certification program.

For network analysis, the easy pick is Gephi. On the download website, the section Learn how to use Gephi (https://gephi.org/users/) contains a brief official tutorial, the documentation and pointers to many other guides, courses and manuals in different languages.

System dynamic and agent-based models' most common development platform is NetLogo. The platform's website has all the online documentation and brief tutorials to start using the software (https://ccl.northwestern.edu/netlogo/docs/tutorial1.html).

Additionally, Marco Janssen's online book: *Introduction to Agent Based Modeling* (https://cbie.gitbook.io/introduction-to-agent-based-modeling/) is based on NetLogo and provides a good introduction to the program and its functionalities as well as a number of worked examples.

Index

For Product Safety Concerns and Information please contact our EU Authorised Representative:

Easy Access System Europe

Mustamäe tee 50

10621 Tallinn

Estonia

gpsr.requests@easproject.com

www.ingramcontent.com/pod-product-compliance
Lightning Source LLC
Chambersburg PA
CBHW071128050326
40690CB00008B/1376